SpringerBriefs in Biology

T0233688

For further volumes:
http://www.springer.com/series/10121

Anthony Reading

Meaningful Information

The Bridge Between Biology, Brain, and Behavior

 Springer

Anthony Reading
University of South Florida
5004 Landstar Way
Florida, 33647, USA
areading@verizon.net

ISBN 978-1-4614-0157-5   e-ISBN 978-1-4614-0158-2
DOI 10.1007/978-1-4614-0158-2
Springer New York Dordrecht Heidelberg London

Library of Congress Control Number: 2011929874

© Springer Science+Business Media, LLC 2011
All rights reserved. This work may not be translated or copied in whole or in part without the written permission of the publisher (Springer Science+Business Media, LLC, 233 Spring Street, New York, NY 10013, USA), except for brief excerpts in connection with reviews or scholarly analysis. Use in connection with any form of information storage and retrieval, electronic adaptation, computer software, or by similar or dissimilar methodology now known or hereafter developed is forbidden.
The use in this publication of trade names, trademarks, service marks, and similar terms, even if they are not identified as such, is not to be taken as an expression of opinion as to whether or not they are subject to proprietary rights.

Printed on acid-free paper

Springer is part of Springer Science+Business Media (www.springer.com)

To

Beth,

Wendy, Sarah, Greg, and Luke

and

William, Caroline, Faye, and Leo

The real solution to the mind-brain-body problem rests in how information is to be conceived in living things, in general, and the brain in particular.

— J. A. S. Kelso, 1995

Preface

The book introduces a radically new way of thinking about information and the central role it plays in living systems. It opens up new avenues for exploring how cells and organisms change and adapt, since the ability to detect and respond to meaningful information is the key that enables them to decipher their genetic heritage, regulate their internal milieu, and navigate safely through their surroundings. Information is a function of the way that matter and energy are organized and arranged in space and time, a property of their *form* rather than their substance. Form refers to the shape and appearance of material objects, like the contour of a palm tree or the outline of the Eiffel Tower. Form and substance are inseparable, however, since it is impossible to conceive of form that has no substance, or substance that has no form. But patterns of matter and energy only convey *meaning* to someone or something that can detect and decipher them—as they otherwise have no effect. Each cell type and each species is endowed with receptors that enable them to perceive a subset of such patterns. The particular patterns they can detect have been shaped by natural selection to enable them to function effectively within their particular ecosystems. Biological detection and response systems range from the chemical configurations that govern genes and cell life to the relatively simple tropisms that guide single-cell organisms, the rudimentary nervous systems of invertebrates, and the complex neuronal structures of mammals and primates.

Meaningful information is defined here as a pattern of matter or energy that is detected by an animate or manufactured receptor that then triggers a change in the detecting entity's behavior, functioning, or structure. The detecting entity can either be a macromolecule, a cell, an organism, a plant, an animal, or a fabricated device; and the change it generates may be either a behavioral one, like fight or flight, a physiological one, like salivating or sweating, or a structural one, like reconfiguring the neural connections involved in learning and memory. Detected patterns of matter or energy that have no effect on a recipient are considered to be meaningless as far as that particular individual is concerned at that particular time. This concept provides a way of understanding how living entities interact with the environment and with each other, separate from the way they interact and respond to physical stimuli. The book thus touches on topics that range from cellular signaling to conscious decision-making, from category formation to goal-directed behavior, from genetic and emotional information to analog and digital forms of representation, and from information theory and neural circuits to maladaptive behavior and the mind–brain interface.

Meaningful information and free energy are both properties of organized matter that function as causal agents, although they do so through different mechanisms. The effects caused by information differ from those caused by energy in that they are primarily determined by the recipient, rather than the initiating entity. The effects that detected information elicits are activated by energy supplied by the recipient and determined by its molecular or neural connections, rather than by the initiating stimulus, so that neither the magnitude nor the type of the response has any relation to the event that triggers it. This is why the way a cell or organism (or device) responds to information cannot be predicted on a purely mechanical basis, why the amount of information being conveyed is independent of the amount of energy used to convey it, and why information-caused events do not follow Newton's laws. This is also why the behavior of living entities cannot be explained entirely in the mechanistic terms used to understand the physical world—and why biology cannot simply be reduced to physics and chemistry.

Both our genetic and experiential heritages are basically informational sets that tell our cells what to do, our organs how to function, and ourselves how, when, where, and why to act. All living creatures exhibit goal-directed behavior, from the bacteria that seek out food sources to the birds that fly south for the winter to the stockbrokers who trade on Wall Street. The inability of physical explanations to account for this type of behavior has been a major obstacle to understanding the mind as a part of the natural world—for how can future outcomes that have not yet happened cause individuals to behave the way they do? It is no wonder that people thought that some sort of mystical process must be involved in causing this, some as yet unidentified vital force or spirit. They were right about there being another process, but not about what it was. We no longer have to equate nonmechanical causation with magic and superstition, since the detection of meaningful information can initiate changes in biological and behavioral systems that are not explainable on a purely physical basis. The vital spirit that animates living things is not some ethereal force beyond our grasp, but simply is the ability of cells and organisms to detect and respond to meaningful information.

The book is thus aimed at a broad audience, primarily in the fields of general and evolutionary biology, cognitive psychology, neuroscience, and philosophy. It explores the larger picture that links these disciplines together, rather than the details that separate them from each other. Although anchored in the discoveries and insights of the particular disciplines, this wider perspective weaves these together to provide new ways of understanding what information is and what it does. However, because this way of thinking about information does not fit neatly into any one of these disciplines, it sails at times against some of their conventional views. Knowledge and understanding advance both through exploring the details of how entities function and by linking these back together to explain why they do. This book follows this latter tradition and is, as a consequence, based more on research in the library than in the laboratory. The many authors on whose ideas and efforts it has been built are gratefully acknowledged in the text and listed in the references.

Contents

Chapter 1
Introduction

Abstract This chapter provides an overview of what lies ahead. It describes how the special uses of the term "information" since the development of Cybernetics and Information Theory have resulted in confusion about its meaning, so that most people are not clear about just what it is that brains and computers process. Meaningful Information is defined as a detectable pattern of matter or energy that generates a response in a recipient. The response may be either a behavioral one, like fight or flight, a physiological one, like salivating or sweating, or a structural one, like reconfiguring the neural connections involved in learning and memory. Meaningful information plays a central role in biological systems, from genes to cells and microorganisms, to multicellular plants and animals.

Information is one of those words that are hard to define—at least in a way that everyone agrees. Even though most of us have a general idea of what the word means, we are not sure whether something we do not understand or already know is information; or something that is not true; or if the library in Beijing contains information for people who do not understand Chinese. It is also not clear whether the smell of a rose or the pattern of stars in the sky represents information; or the evening news, the data stored in a computer, or a pain in your big toe; or rain clouds in the sky and footprints in the snow; or the tree that falls in the forest when no one is around; or even the things that do not actually happen, like Sherlock Holmes' dog that did not bark. It seems that no one meaning is correct, since all of these examples are included in some definitions, although not in others.

Despite the fact that information is such an inescapable part of modern life, most people have trouble explaining exactly what it is or what it does. Like the smile on the proverbial Cheshire cat, it is strangely elusive when we examine it closely. We know it is something that brains and computers process, yet there is no trace of it when we look inside them or take them apart. The reason is that information is not an object like the neurons and semiconductors that convey it, but is a function of the

A. Reading, *Meaningful Information: The Bridge Between Biology, Brain, and Behavior*, SpringerBriefs in Biology 1, DOI 10.1007/978-1-4614-0158-2_1, © Springer Science+Business Media, LLC 2011

way these are arranged. Information is an intangible concept that does not have anything to anchor it to material reality, so that different individuals and different disciplines are free to use it in ways that suit their particular purposes. The situation is a little like *Through the Looking-Glass,* where Humpty Dumpty observes: "When *I* use a word, it means just what I choose it to mean—neither more nor less," to which Alice replies: "The question is whether you *can* make words mean so many different things," and Humpty responds: "The question is which is to be master— that's all" (Carrol 1872, 1996). However, this diversity can lead to misunderstandings and mistakes when people with different ways of defining information want to communicate about it, which is why the way the word is used in scholarly endeavors needs to be clearly specified.

The Source of Confusion

Information's everyday meaning in English refers to something that *informs* someone, that is, to knowledge communicated about a particular fact, subject, or event. Most of the current confusion about its meaning began in the middle of the twentieth century, when the term was applied to phenomena in cybernetics, computers, and data transmission. These new frontiers expanded its use to include aspects of the universe that did not communicate or affect anything. Whereas information's ordinary meaning depended on having an observer perceive what was being communicated, these new uses envisioned it as a freestanding entity that existed independent of being observed. The two most influential contributors to the new ways of thinking were Norbert Wiener (1894–1964) and Claude Shannon (1916–2001), both of whom were giants of the modern Information Age. In 1948, Wiener introduced the concept of information as a measure of the degree of organization in a system and, in 1964, Shannon unveiled the concept of information as something that reduces uncertainty in a message.

Wiener, who was the founder of the field of cybernetics, saw information as an intrinsic property of all organized forms of matter and energy, something that reflected the degree to which the way their components were organized differed from randomness. The way the letters and words on this page are arranged, for instance, represents the information they convey, since there would not be any if they were set out randomly. As he put it, "Just as the amount of information in a system is a measure of its degree of organization, so the entropy of a system is a measure of its degree of disorganization; and the one is simply the negative of the other" (Wiener 1948, p. 18). Wiener's concept is a theoretical one that applies to everything in the known universe, whether or not it has any detectable meaning or effect. Only a small fraction of the way the universe is organized is accessible to us, however, so that we have no meaningful way of detecting this type of information in most naturally occurring objects, like trees or ant colonies or brains.

Shannon introduced a mathematical concept of information as part of the *Information Theory* he developed and demonstrated how this could be measured by the amount of uncertainty a transmitted message reduced (Shannon and Weaver 1964). His theory deals, however, with the *capacity* of a system to transmit information, and not with the content or meaning of the information being transmitted. As a result, two completely different messages, one loaded with meaning and the other absolute nonsense, can contain exactly the same amount of Shannon-type information. Even though Shannon stated he was using the term "information" in a special sense that should not be confused with its ordinary meaning, and actually called his proposal a communication theory, its name gradually changed as its influence became more widespread.

The use of the word then increased exponentially as computers proliferated, the genetic code was broken, and the Internet grew and flourished. A variety of new disciplines were even created around it, including Information Science, Bioinformatics, and Information Technology, although none of them helped clarify its meaning (Young 1987).[1] We were, as a result, transformed into an *Information Society* in which science and technology focused increasingly on information-related topics. Rather than clarify matters, these developments simply converted information into one of those multipurpose words that can refer to a number of things. Amazon.com, for instance, currently lists over 125,000 books that include the word in their title, although many of them have little else in common. As Schiller (2007, p. 17) notes: "The study of information is uncomfortably disparate, since the term refers to diverse phenomena in library and information science, engineering, communication, sociology, law, economics, literary criticism, history, and other fields."

The various ways information is now conceptualized depend mainly on whether it is seen as a tangible and potentially measurable aspect of the universe or as an intangible and nonquantifiable feature of the form that matter and energy take. As Edelman and Tononi (2000, p. 209) note: "Inquiry about the origin of information in nature faces several definitional problems, many of which are concerned with the distinction between the describer and the thing described. First, we must ask whether the term *information* can be used to describe a state of nature in the total absence of a human observer. Can *information* be solely an objective term? If it is defined by physicists as a measure of order in a far-from-equilibrium state, then by the terms of that definition, in a 'God's eye view,' it is objective. If, however, information is defined in a way that requires a historical process involving either memory or a heritable state, then information can only have arisen with the origin of life."[2]

Conveying Meaning

In spite of the numerous advances that have been made in processing, storing, and transmitting information, its meaning remains elusive. The problem is that *meaning* is not an integral property of objects and events, but something that the observer attributes to them. Meaningful information is thus something that exists solely in the eyes of the beholder—which is why different individuals and different species can attribute different meanings to the same information. The entire spectrum of living entities, from the tiniest cells and organisms to the grandest plants and animals, is able to detect and respond to information that is meaningful, as also are some of the devices our species have invented. For information to convey meaning, however, it not only has to be detected but also has to have an effect on the entity that detects it. Every species is comprised of cells and molecules that are specifically designed to detect and respond to the types of information it needs to function and adapt in its particular environment. Information that has no detectable effect on anything is, for all practical purposes, devoid of any meaning (Day 2001).

A number of attempts have been made to recognize the way that certain aspects of information have an effect on the recipients that detect them (i.e., are meaningful to them), although none has met with wide acceptance. MacKay (1969), for instance, defined information as a relationship between a message and a recipient, Gatlin (1972) as the capacity to store and transmit meaning or knowledge, Bateson (1979) as a difference that makes a difference, Stonier (1997) as possessing the capacity to organize, Dretske (2000) as an event that makes a causal difference, and Jablonka and Lamb (2005) as something that changes the recipient's functional state. Although each of these captures some facet of the phenomenon, none does so entirely. Rather than offer another general way of defining information, this book focuses on the aspect of it that can elicit a response in the entities that detect it.

The term *Meaningful Information* is thus used throughout the book to refer specifically to a detected pattern of matter or energy that generates a response in a recipient. The response may be either a behavioral one, like fight or flight, a physiological one, like salivating or sweating, or a structural one, like reconfiguring the neural connections involved in learning and memory. If a detected pattern of matter or energy has no effect on a recipient, it is considered to be meaningless as far as that individual is concerned. This way of thinking follows the pragmatic approach of Charles Saunders Peirce and William James, both of whom argued that things should be defined in terms of the effects they produce.[3] James (1907, p. 46) quotes Peirce's belief that "to develop a thought's meaning, we need only determine what conduct it is fitted to produce: that conduct is for us its sole significance." Meaningful information, as here defined, is thus similar in some ways to Wiener's (1967, p. 127) concept of *semantically significant information* and Roederer's (2005, p. 4) notion of *pragmatic information,* which: "must not only have a purpose on the part of the sender, it must also have a meaning for the recipient in order to elicit the desired change."

Because meaning is seen here primarily as a function of the observer, rather than of what is being observed, information that is meaningful to one individual, one species, one machine, or one cell may not be meaningful to another. It even may not be meaningful in the same way to a given individual at a different time or in a different context. Although the type of information that cells, plants, and invertebrates can detect is primarily specified by their genes, learning and experience also shape much of what the other species are able to detect and respond to. In our own one, for instance, the different strata in rocks and the different images on X-rays only represent meaningful information to individuals who are trained in these particular fields. While we are conscious of the meaning we attribute to the information our sensory system detects, species that lack consciousness respond automatically to the information they sense, without knowing what they are doing or why they are doing it. The pattern of light flashes emitted by fireflies, for instance, automatically conveys meaningful information to other fireflies, the wiggle dance of bees to other bees, and the mating call of frogs to other frogs. We infer that these signals represent meaningful information because of the obvious responses they elicit, rather than by understanding what the recipients actually experience. Meaning, as defined here, is something that is embedded in the response of the receiver, rather than a property of the objects and events that elicit it.

Meaningless information thus consists of patterns of matter and energy that either cannot be detected or, although detected, have no effect on the recipient. The reason that so much of the information in the universe appears to be meaningless is that only specific kinds of devices can detect it. These consist of the specialized sensory and molecular receptors with which living cells and organisms are endowed, as well as various fabricated devices people have invented. Since the vast majority of the patterns of matter and energy in the universe are beyond the capacities of known detectors, their meaning remains unknowable. Many of the information patterns that are capable of having an effect on some kind of recipient are, however, not currently doing so, including most of the material stored in libraries, archives, and the Internet. These potentially meaningful forms of information are referred to here as *data*. Data represent inactive forms of information that can be detected and have an effect on an entity, but are not currently doing so. Information can thus be classified in roughly the same way as energy: either as a fundamental property of organized matter that has no discernable effect (intrinsic information and internal energy), an inactive subset of this that can have an effect, but is not currently doing so (data and potential energy), or an active one that is currently having an effect (meaningful information and free energy). This is no coincidence, since energy and information are closely related constructs.

Cause and Effect

Information is taken here to be an expression of the way that matter and energy are organized in space and time, a property of their *form* rather than their substance. Informational patterns, however, only convey *meaning* to someone or something that can detect and respond to them, either by changing their behavior, physiology, or neural organization. The way that matter and energy are arranged in the universe is like a secret code, so that only the small fraction of it that can elicit a response in some kind of recipient is considered to be meaningful. Although, as Wiener proposed, information is theoretically a property of all organized forms of matter and energy, only the small fraction of it can elicit a response in some kind of recipient is considered to be meaningful. This latter portion is the one that living entities use to regulate their growth, maintain their homeostasis, and adapt to their environment. It is also the one that growing children search for in trying to make sense of the world, that scientists and explorers set out to discover, and that we all consider in trying to understand what we experience, for it is the only aspect of information that conveys meaning.

Meaningful information and free energy are both envisioned here as aspects of organized matter that are able to bring about change, although they do so through different mechanisms. The effects caused by information differ from those caused by energy in that they are primarily determined by the recipient, rather than the initiating entity. Neither the magnitude nor the type of response to detected information has any necessary relation to the object or event that triggers it, since (a) the effects it elicits are activated by energy supplied by the recipient, rather than the initiating stimulus, and (b) the nature of the response is determined by the recipient's molecular and neural connections, rather than the physical properties of the initiating stimulus. Merely pushing a button, for instance, can transmit meaningful information that can either ring a doorbell or launch a rocket to the moon, based on how the entity that detects the signal is configured. This is why the nature of the response to information cannot be predicted on a purely physical basis, since information-caused events do not follow Newton's laws. It is also why both the amount of information being conveyed and the magnitude of the effect it triggers are independent of the amount of energy used to convey it. Although energy and information both function as change agents in the animate world, information has no effect on the inanimate objects that exist in nature. This is why the behavior of living entities cannot be explained entirely in the mechanistic terms used to understand the physical world—and why biology cannot simply be reduced to physics and chemistry.

Ever since Descartes (1596–1650), philosophers have wondered if the workings of the brain could be explained in mechanical terms, or whether some mysterious type of "vital force" was involved. Descartes held that although the body functioned mechanically, the mind was a nonphysical entity that acted separately from it to produce conscious thought and self-awareness, but he was unable to explain how a nonmaterial entity could cause material effects. On the other hand, scholars who reject his supernatural ideas and believe the brain can be understood entirely in physical terms have not yet found a way of explaining the mental phenomena it produces. The concept of meaningful information offers a bridge between these two points of view, since it represents a nonmechanical form of causation that is not tainted by spiritual and metaphysical beliefs. As Nadel (1979, p. 431) notes: "There is a genuine alternative in biology to both vitalism and mechanism—namely the development of systems of explanation that employ concepts and assert relations neither defined in nor derived from the physical sciences." Information and energy intermingle as causal agents in biological systems, with one generating responses to "form" and the other responses to "substance." Form refers to the shape and appearance of material objects, like the contour of a palm tree or the outline of the Eiffel Tower. Form and substance are inseparable, however, since it is impossible to conceive of form that has no substance or substance that has no form. Thus, although information requires a material medium to portray and communicate it, the medium is not the message.[4]

Notes

[1] *Information Science* is an interdisciplinary field that deals primarily with the collection, classification, manipulation, storage, retrieval, and dissemination of accumulated information. *Bioinformatics* deals with the use of mathematical, statistical, and computing methods to study the role of biomolecules in cellular regulation. *Information Technology* is a branch of electrical engineering and applied mathematics that deals with the study, design, development, implementation, support, and management of computer-based information systems, based on Shannon's Information Theory (which is discussed further in Chap. 17).

[2] Floridi (2010) provides a useful overview of the conventional ways of understanding information as an objective and potentially measurable aspect of the universe. These are based primarily on concepts derived from Shannon-type theories and computer technology—and do not deal with the more subjective aspects of the topic. Nauta (1972) attempts to integrate Shannon's Information Theory and Semiotics (the study of signs) into a coherent explanatory theory but fails to overcome the problems that both face in explaining the "subjective" aspects of biological information. None of the theories that envision information as an objective phenomenon can explain how it can cause something to happen, like how a red light can cause a car to stop, or how different individuals can respond in different ways to the same information.

[3] Peirce (1838–1914) is generally credited with being the founder of Pragmatism. James (1842–1910), the great pioneer of American psychology, befriended Peirce and helped spread his philosophical ideas. Peirce also established the field of Semiotics, the first discipline to study information in a systematic way. Short (2007) provides an account of Peirce's various contributions to philosophy.

[4] Rene Descartes (1596–1650) proposed that there are two fundamental kinds of substance in the universe: material things (res extensa) that obey the laws of physics, and thinking and spiritual ones (res cogitans) that do not. He thought that the body was composed of material components that worked together like a machine, while the mind was a nonphysical entity that controlled the body. This way of separating the mind and the body remained the dominant framework for understanding human behavior for almost 350 years.

References

Bateson G (1979) Mind and Nature: A Necessary Unity. Dutton, New York

Carrol L (1872, 1996) Through the Looking-Glass. Puffin Books, New York

Day RE (2001) The Modern Invention of Information. Southern Illinois University Press, Carbondale

Dretske FI (2000) Perception, Knowledge, and Belief: Selected Essays. Cambridge University Press, Cambridge

Edelman GM, Tononi G (2000) A Universe of Consciousness: How Matter Becomes Imagination. Basic Books, New York

Floridi L (2010) Information: A Very Short Introduction. Oxford University Press, Oxford

Gatlin LL (1972) Information Theory and the Living System. Columbia University Press, New York

Jablonka E, Lamb M (2005) Evolution in Four Dimensions: Genetic, Epigenetic, Behavioral, and Symbolic Variation in the History of Life. MIT Press, Cambridge

James W (1907) Pragmatism: A New Name for Some Old Ways of Thinking. Longmans, Green & Co, New York

MacKay DM (1969) Information, Mechanism and Meaning. MIT Press, Cambridge

Nadel E (1979) The Structure of Science. Hackett Publishing, Indianapolis

Nauta D Jr (1972) The Meaning of Information. Mouton, The Hague

Roederer J G (2005) Information and Its Role in Nature. Springer, New York

Schiller D (2007) How to Think About Information. University of Illinois Press, Urbana

Shannon C, Weaver W (1964) The Mathematical Theory of Communication. The University of
   Illinois Press, Urbana
Short TL (2007) Peirce's Theory of Signs. Cambridge University Press, New York
Stonier T (1997) Information and Meaning: An Evolutionary Perspective. Springer-Verlag,
   London
Wiener N (1948) Cybernetics: or Control and Communication in the Animal and the Machine.
   John Wiley & Sons, New York
Wiener N (1967) The Human Use of Human Beings. Avon Books, New York
Young P (1987) The Nature of Information. Praeger, New York

Chapter 2
Meaningful Information

Abstract "Meaningful Information" is defined as a pattern of organized matter or energy that is detected by an animate or manufactured receptor and thereby triggers a change in the behavior, functioning, or organizational structure of the detecting entity—which may either be a macromolecule, a cell, an organism, a plant, an animal, or a fabricated device. A great deal of what is currently called "information" does not fit this definition, since it does not change or affect the recipients. The ability to detect and respond to meaningful information is essentially a biological phenomenon, since there are no inanimate information detectors in nature. Information and energy are both fundamental properties of organized matter that reflect the complexity of its organization, but while energy is a function of an entity's mass or substance, information is a function of its form (i.e., of the way its structure is organized and arranged in space or time).

The problem with offering yet another meaning to a word that already has several is not that people will not be able to understand it, but that they already think they do. The concept of meaningful information presented here represents a new way of looking at how cells and other living entities communicate with each other and how they detect, process, store, and respond to certain patterns of matter and energy. Its impact extends far beyond the word's associations with knowledge, language, and computers. Meaningful information is something that lights up the entire biosphere with a constant chatter of cellular signals that orchestrate how living organisms grow, adapt, and communicate with each other. It is the type of information that Oyama (2000, p. 1) refers to when she comments: "Information is what enables molecules, cells, and other entities to recognize, select, and instruct each other, to construct others and themselves, to regulate, control, induce, direct, and determine events of all kinds."

The basic idea is fairly simple. Meaningful information consists of patterns of matter or energy that have an effect of some kind on the entities that detect them—with the effect being either a behavioral response, a physiological change, or an

A. Reading, *Meaningful Information: The Bridge Between Biology, Brain, and Behavior*,
SpringerBriefs in Biology 1, DOI 10.1007/978-1-4614-0158-2_2,
© Springer Science+Business Media, LLC 2011

alternation in their neuronal connectivity. A great deal of what is currently called "information" does not fit this definition, since it does not change or affect the recipients. Although their sensory systems can detect the patterns being conveyed, their brains cannot discern any meaningful information in them. This is because meaningful information, at least as envisioned here, is not a property of the patterns of matter or energy themselves, but something attributed to them by the recipient. The arrangement of the stars in the sky or the mating call of a Carolina Wren, for instance, represent meaningful information only to certain individuals, but not to others, since it has no effect on them. Meaningful information does not even have to be true; it just has to have an effect on the recipients—although they usually have to believe it is true in order to respond. Even having a sender is not essential, since naturally occurring phenomena can convey meaningful information without one, like fossils in a rock or the position of the sun in the sky.

Key Concepts

This book is about a particular way of conceptualizing information, the key elements of which are described below. As Gatlin (1972, p. 25) notes: "Information is an ultimately indefinable or intuitive first principle, like energy, whose precise definition always seems to slip through our fingers like a shadow. This does not mean that we cannot define information *operationally* as we do energy and understand a great deal about its nature and how it expresses itself in the world about us." This is why meaningful information is defined here by what it *does*, rather than what it is.[1]

Intrinsic information is a theoretical concept proposed by Wiener (1948) that refers to the way the various particles, atoms, molecules, and objects in the universe are organized and arranged. The amount of intrinsic information an entity contains is a measure of its organizational complexity, an expression of the degree to which the elements that comprise it are arranged in a nonrandom fashion. Although the amount of intrinsic information can be quantified in man-made devices that store or transmit data, it cannot be measured directly in naturally occurring entities, since we have no way of accessing it in them. As Goldstein and Goldstein (1993, p. 133) explain: "we can measure any changes of energy or information a body undergoes, but not how much energy or information there was in the body in the first place—since there are no instruments that can measure them, no energy-meters or entropy-scopes."

Meaningful information refers here to the minute fraction of intrinsic information that can be detected and can cause a change in some sort of recipient. It is defined operationally as a spatial or temporal pattern of organized matter or energy that is detected by an animate or manufactured receptor, which then triggers a change in the behavior, functioning, or structure of the detecting entity. The detecting entity can either be a macromolecule, a cell, an organism, a plant, an animal, or a fabricated device. If there is no effect on the detecting entity's behavior, functioning, or structure, the information is considered to be meaningless as far as that particular individual is concerned at that particular time. Meaning is the key that enables information to have an effect, the attribute that transforms detected patterns of matter and energy into signs, signals, and messages that *inform* the recipient. Patterns of organized matter or energy that *can* have an effect on an animate or human-made receptor, but are not currently doing so, are referred to here as *data*. They include the material stored in computer memories and libraries, the initial results of experiments, and messages in codes and foreign languages. Arrangements of matter or energy that can be detected but are not capable of generating an effect or change in a recipient are referred to here as *noise*.

Biological Information

Meaningful information is essentially a biological phenomenon, for there are no inanimate information detectors in nature. Although some fabricated devices can detect and transmit information, they are unable to do so on their own, since someone has to design them, someone has to encode them, and someone has to interpret their output. While both living and nonliving entities respond to the physical aspects of matter and energy, only living ones detect and respond to the *form* of the objects and events they constitute. The ability to detect and respond to information is, in fact, one of the defining attributes of living things, something that is lost when they eventually die. Form plays a central role in all of biology, since it is also expressed in the shapes of organisms, the arrangement of their cells, and the folding pattern of the protein molecules these contain, all of which serve adaptive functions that have been shaped by natural selection over the millennia. As Mayr (1982, p. 52) notes: "The explanatory equipment of the physical sciences is insufficient to explain complex living systems and, in particular, the interplay between historically acquired information and the responses of these genetic programs to the physical world. The phenomena of life have a much broader scope than the relatively simple phenomena dealt with by physics and chemistry."[2]

Dusenbery (1992, p. 19) points out, "Organisms obtain information about their environment to help solve a wide variety of problems—maintaining an appropriate environment, timing activities, locating resources or threats—and many organisms transmit information to other individuals in order to persuade them to do something." All living entities import information from the outside world to enable them to learn and adapt to their surroundings in much the same way as they import energy to fuel their various activities. While cells and unicellular organisms respond automatically to information patterns that are meaningful to them, more complex creatures develop neuronal structures that enable them to store and interpret what they perceive, and to vary their responses accordingly. According to Gatlin (1972, p. 1): "Life may be defined operationally as an information processing system—a structural hierarchy of functioning units—that has acquired through evolution the ability to store and process information necessary for its own accurate reproduction."

While *order* is a general property of nature, *organization* is an exclusive manifestation of living things, ranging across the entire spectrum from cells to organs to individuals, as well as to the devices and social systems they create. Order simply indicates that entities are arranged in a nonrandom fashion, like the furniture in a room, the bases in a strand of DNA, or the steel girders in a skyscraper. Organization, on the other hand, implies that entities are not only arranged in an orderly manner but that their component parts also interact with each other in ways that affect the entity's overall functioning. The essence of organization is that the entity's components communicate with one another by sending and receiving meaningful information. This is what enables them to coordinate their activities and function as an *ensemble*, rather than just a collection of parts. As Harold (2001, p. 108) notes: "Living things differ from non-living ones most pointedly in their capacity to maintain, reproduce, and multiply states of matter characterized by an extreme degree of organization."

Information and Energy

Energy is defined as the capacity to do work or cause physical change. The modern concept of energy was introduced in the eighteenth century in association with the emerging science of thermodynamics. As Dear (2006, p. 16) points out, it shares a number of attributes with information: "Energy was a very abstract concept, with no pure form that could be put in a bottle on the laboratory shelf, but its use as an accounting procedure for tracking changes in physical phenomena quickly made it indispensable." Virtually all of the available energy on our planet is derived from the sun. Plants capture energy from it through photosynthesis and store it for later use. Animals obtain their supply of energy by metabolizing plant substances (i.e., breaking down their organized structure), either directly or indirectly. Fossilized plant and animal matter, such as coal and oil, can also be broken down into less-complex substances when oxidized, thereby releasing energy that can be harnessed for work (Bryant et al. 2006).[3]

Information and energy are both fundamental properties of organized matter that express the complexity of its organization. But whereas energy is a function of an entity's *mass* or *substance*, information is a function of its *form*, of the way it is arranged in space and time. An entity with a repetitive pattern of structural elements, such as a crystal or a lattice, contains less information than a more intricately organized one, but not less energy. A 10-lb sack of potatoes, for instance, contains twice as much energy as a 5-lb one, but not twice as much information; nor do a dozen copies of the *Encyclopedia Britannica* contain more information than a single one, since the additional copies are just redundant. Because of the way redundancy reduces the amount of information in an entity, but not the amount of energy, the quantity of information a given object contains is not necessarily the same as the quantity of energy.[4]

The information and energy that are present in organized matter are only inactive potentials, however, until some process enables them to be realized. Most of the physical entities that make up the universe, such as atoms, mountains, and stars, are relatively stable and cannot easily be broken down to reveal either the energy or the information they contain. For energy to become available, an organized entity has to be transformed into a less-organized state, which is what happens when fossil fuel is burned or a nuclear reaction takes place. For information to become available, however, the object must interact with a receptor (or a set of receptors) that is capable of detecting the meaning it conveys. Thus, although the amount of energy in a source becomes depleted when some of it is released, the amount of information it contains is not changed by being detected. The words on this page, for instance, do not become degraded as you read them, but remain intact for others to read later on. Moreover, while energy cannot be destroyed, information can, as by burning a book, accidentally erasing a computer disc, or getting Alzheimer's disease.

Because a given amount of matter can exist in a number of different forms, the information each of them contains is not the same. Thus, although the arrows below contain the same amount of matter, they do not convey the same information. On the other hand, the squares adjacent to them convey the same information, even though they consist of different amounts of matter.

The amount of energy used to transmit information is unrelated to the amount of information being transmitted, which is why different energy substrates can be used to convey the same information—as long as they have comparable spatial and temporal patterns. This is the basis of the process of *transduction*, in which an information pattern in one energy medium is converted into an equivalent pattern in another. Transduction is the mechanism that enables electronic signals to be converted into radio and television shows, and sensory information to be represented in the brain as neuronal configurations. The key distinctions between energy and information are captured by Von Bayer (2004, p. 11): "Energy is located strictly within a physical system—metabolic energy in a jelly doughnut, electrical energy in a mobile-phone battery, chemical energy in a gas tank, kinetic energy in every puff of wind. Information, on the other hand, resides partly in the mind. A coded message may represent gibberish to one person, and valuable information to another. The smell of subjectivity, of dependence on a state of mind is the source of both the elusiveness and the power of the concept of information." This is why, as he sees it, "information stands poised to join the concept of *energy* as a unifying thread that runs through all of science, linking diverging branches and unifying the foundations of the whole enterprise" (p. 99). But first we need to be clear which sort of information we are talking about.[5]

Notes

[1] The everyday meaning of *information* emerged in the fourteenth century from roots in the Latin *informatio*, meaning to form, place in order, or arrange. It was something that gave *form* to the mind. Dr. Johnson's 1755 dictionary described it as the process of telling something to someone

and to the content being transmitted. The 1961 *Oxford English Dictionary* defined it as: (a) the action of informing, formation or molding of the mind or character, training, instruction, teaching, (b) the act of informing; communication of the knowledge or "news" of some fact or occurrence; the act of telling or the fact of being told of something, and (c) knowledge communicated concerning some particular fact, subject, or event; that of which one is appraised or told; intelligence, news (Cappuro 1985).

[2] Maynard Smith (2000) provides a lucid review of the unique role that information plays in the evolution and maintenance of biological systems, while Godfrey-Smith (2007) reviews the philosophical issues involved in the use of informational concepts in biology.

[3] The energy derived from nuclear reactions is an exception.

[4] The amounts of information and energy an entity contains are not necessarily identical, even though both are functions of the degree by which its organizational structure differs from randomness. The two of them thus entail somewhat different concepts of entropy, since this is an inverse measure of the amount of order and organization. *Entropy* refers to both the amount of disorder in a thermodynamic system as well as the process by which disorder occurs. According to the second law of thermodynamics, the total entropy of an isolated thermodynamic system tends to increase over time, so that organized systems tend to run down and become less organized unless external energy is supplied to them. This is why hot objects become cold, physical structures decay, and living things need to import energy to stay alive. As La Cerra (2003) notes: "Entropy is the primary adaptive problem that life forms must solve; consequently, evolutionary processes have crafted intelligence systems that are fundamentally designed to acquire, manage and direct energetic resources toward the maintenance of life processes and the attainment of life-stage specific goals." A review of the differing concepts of entropy can be found at http://entropysite.oxy.edu.

[5] Gitt (1996, p. 181) maintains that: "Energy and matter are considered to be basic universal quantities. However, the concept of information has become just as fundamental and far-reaching, justifying its categorization as the third fundamental quantity." As he sees it (Gitt 2006, p. 121): "Although matter and energy are necessary fundamental properties of life, they do not in themselves imply any basic differentiation between animate and inanimate systems."

References

Bryant JA, Atherton MA, Collins MW (2006) Design and Information in Biology: From Molecules to Systems. WIT Press, Southampton

Cappuro R (1985) Epistemology and Information Science. Report Trita-Lib-6023, S Schwarz (Ed). Available at http://www.capurro.de/trita.htm

Dear P (2006) The Intelligibility of Nature: How Science Makes Sense of the World. University of Chicago Press, Chicago

Dusenbery DB (1992) Sensory Ecology: How Organisms Acquire and Respond to Information. WH Freeman, New York

Gatlin LL (1972) Information Theory and the Living System. Columbia University Press, New York

Gitt W (1996) Information, science and biology. J. Creation 10(2): 181–187

Gitt W (2006) In the Beginning Was Information: A Scientist Explains the Incredible Design in Nature. Master Books, Green Forest

Godfrey-Smith P (2007) Information in biology. In DJ Hull & M Ruse (Eds), The Cambridge Companion to the Philosophy of Biology. Cambridge University Press, New York

Goldstein M, Goldstein JF (1993) The Refrigerator and the Universe: Understanding the Laws of Energy. Harvard University Press, Cambridge

Harold FM (2001) The Way of the Cell: Molecules, Organisms and the Order of Life. Oxford University Press, New York

La Cerra P (2003) The first law of Psychology is the second law of Thermodynamics: The energetic evolutionary model of the mind and the generation of human psychological phenomena. Human Nature Review 3: 440–447

Maynard Smith J (2000) The concept of information in biology. Philosophy of Science 67: 177–194

Mayr E (1982) The Growth of Biological Thought. Harvard University Press, Cambridge

Oyama S (2000) The Ontogeny of Information: Developmental Systems and Evolution, 2nd edn. Duke University Press, Durham

Von Bayer HC (2004) Information: The New Language of Science. Harvard University Press, Cambridge

Wiener N (1948) Cybernetics: or Control and Communication in the Animal and the Machine. John Wiley & Sons, New York

Chapter 3
Cause and Effect

Abstract Information and energy are the two fundamental agents of change in the natural world, although there are critical differences in how they operate. The energy involved in physical causation is supplied by the originating entity, while the energy involved in informational causation is supplied by the receiving one. The response to information is also determined primarily by the recipient, not the information that generates it—since the latter is just an on/off signal in most biologic systems. This is why physical explanations have not been as helpful in understanding biological and behavioral phenomena as they have been in understanding physical and chemical ones. Biological systems are exquisitely self-regulating and make extensive use of error-correcting feedback information to maintain their internal environment within certain limits, as well as to regulate their interactions with the external one. The inability of materialistic explanations to account for goal-directed behavior and historical forms of causation has been a major obstacle to understanding the mind as a part of the natural world.

Information and energy are not tangible entities, like chairs or tables; they are simply constructs that have been developed to explain the causal interactions that take place between material things (Von Bayer 2004, p. 130). They are the two fundamental agents of change in the natural world, the two basic mechanisms that link cause and effect together. There is, however, a critical difference in how they function: while the energy involved in physical causation is supplied by the originating entity, the energy involved in informational causes is supplied by the receiving one. This is one of the reasons that cells and organisms (and informational devices) have to have an independent supply of energy, and why entities that do not are unable to detect and respond to information. The response to someone shouting "fire" in a crowded theater, for example, is not fueled by the energy used to convey the information, but by energy supplied by the recipients. This is why the outcomes brought about by the detection of information cannot be explained in purely mechanical terms, for there is no direct relationship between the properties of the stimulus and the response it engenders.[1]

A. Reading, *Meaningful Information: The Bridge Between Biology, Brain, and Behavior*, SpringerBriefs in Biology 1, DOI 10.1007/978-1-4614-0158-2_3, © Springer Science+Business Media, LLC 2011

Meaningful information is the means by which one entity can cause a change in the behavior, functioning, or structure of another without a direct exchange of energy or matter between them (Roederer 2005, p. 126). The response to information depends on the recipient having specialized receptors that are able to recognize and respond to certain configurations of energy and matter in its surroundings. The reaction this sets off in the recipient is determined, however, by the way its receptors and effectors are connected, rather than by the detected information itself. The holes in the paper role of a player piano, for instance, are just signals that activate preprogrammed responses, not the physical cause of the music they generate. Because the effects caused by the detection of information are determined by the recipients, rather than the initiators, they cannot be predicted by the usual laws of physics. However, while information can be a necessary cause, it is never a sufficient one, since it cannot produce an effect on its own.[2] Although a predetermined signal can send a spaceship into orbit, the energy and design for this have to be programmed into the spaceship, not the signal. Information and energy nevertheless work hand-in-hand in the biological world to mold the behavior and functioning of every living cell and organism. Since informational causes do not involve mechanical processes, their interactions cannot be expressed quantitatively or represented in mathematical terms.[3]

Explaining Change

Studies with children show that the human brain is programmed to infer cause and effect relationships when one event is perceived to regularly follow another (Piaget and Garcia 1980). This enables growing children to develop models of how things around them work and to use these to gain a degree of mastery over what they encounter. Our ancestors gradually came to understand a great deal about the universe by developing such cause and effect explanations of how things changed and functioned. The process was not foolproof, however, since it can also lead to false beliefs and superstitions, like those in rain-making dances and astrology. Science itself evolved as a more systematic way of understanding causal interactions, both under natural and experimental conditions (Changeux 2004). As scientific knowledge grew, physical explanations about the universe came to replace magical ones, generally because they proved better at predicting important events and occurrences. Mechanistic explanations of natural phenomena have, however, come to dominate much of modern-day science, as if they were the only alternative to supernatural and mystical beliefs. As Rosen (1991, p.23) observes: "To suggest that living things can be explainable in any way other than by the machine metaphor is regarded as unscientific and viewed with the greatest hostility as an attempt to take biology back to metaphysics."[4]

Mechanistic concepts have not been as helpful in understanding biological and behavioral phenomena as they have in understanding physical and chemical ones. As Harold (2001, p. 219) notes: "For three hundred years now, scientists have perceived the world though the eyes of Descartes and Newton: a universe of particles moving in fields of force, whose behavior is fully determined by the overarching laws of physics." However, understanding the way meaningful information can act as a causal agent enables us to move beyond such dualistic ways of thinking. We no longer have to see ourselves as comprised of a material body that obeys the laws of physics and a nonmaterial spirit that is too elusive to study. Meaningful information can generate physical changes in biological and behavioral systems that do not rely on magical or superstitious explanations. The *vital spirit* that animates living things is not some ethereal force beyond our comprehension but is simply the ability of cells and organisms to detect and respond to meaningful information. Our genetic and experiential heritages are basically informational sets that tell our cells what to do, our organs how to function, and our selves how to act.

Causal relationships have three basic elements: (a) the cause must precede the effect, (b) the cause and the effect must be linked by a chain of contiguous events, and (c) the event would not occur without the cause (Dowe 2000). Physical and informational causes satisfy these criteria in different ways. While the links connecting physical causes and effects involve physical chains of events, the links connecting informational ones entail the perception of patterns in the environment, the nervous system, or the genome. Historical processes, like natural selection and memory, etch events into organisms as particular patterns of DNA or neuronal connectivity that can be activated as causal agents at a future time. Informational causes can thus have delayed effects, since the patterns they generate in a cell or organism can continue to exist long after they have been laid down. Soldiers in combat, for instance, can receive both physical and psychological injuries, with the former causing immediate effects and the latter delayed ones that are triggered by persisting memories of the trauma. Because of the time lags involved in historical causes, some scientists dismiss them as examples of "action at a distance"—a phenomenon that characterizes magical beliefs rather than scientific facts. However, unlike supernatural causes, historical ones are connected to the effects they produce by a chain of informational linkages that can be studied and understood.[5]

As Mayr (2004, p. 89) points out: "Perhaps the most profound difference between the inanimate world of the physicist and the living world of the biologist is the dual causation of all organisms. Anything and everything that happens in the physical world is exclusively controlled by the natural laws, and even living organisms and their parts are, as matter, as subject as inanimate matter to these laws. However, organisms are subject to a second set of causal factors, the information provided by their genetic program. The structural laws and the messages from the genetic program function simultaneously and in harmony, but genetic programs occur only in living organisms. They provide an absolute borderline between the inanimate and the living world." Genetic causes are, however, not the only ones mediated by informational linkages, since mental causes are also informational in nature, as effects can be generated by thoughts about the past, the present, or the future.

Mind and Brain

One of the most puzzling challenges in neuroscience is conceptualizing how the mind and the brain interact. How can a diverse assortment of sensory impulses end up as a set of abstract concepts and memories? How can thoughts and desires that have no physical mass or energy cause tangible, real-world effects? How can subjective entities, like feelings and beliefs, cause a collection of physical neurons to govern how we behave? The fact that living entities are able to respond to information, rather than just to matter and energy, provides a way of resolving the dilemma. Information is present in the brain as patterns of neuronal connectivity, so that neural systems are able to detect and respond to these internal configurations in the same way they detect and respond to external patterns of matter and energy. Although we do not know how our thoughts and expectations are represented in the brain, or how they actually cause events to happen, the fact that information-based patterns are capable of activating brain responses provides a nonmystical way of explaining them. As Haught (2006, p. 68) observes: "If information belongs to a realm distinct from sheer materiality, then science itself has now made room for non-material forms of causation in the natural world."

The human brain is endowed with the capacity to process both sensory and symbolic information. A symbol is something that arbitrarily signifies something else, represents it uniquely, and can generate a response that closely resembles the one generated by the thing it represents. Words and numbers are symbols, as also are wedding rings and national anthems. Even though they are physically unlike the entities they represent, symbols can be just as powerful as them in causing change, since they convey the same information. As Dretske (2000) points out, even a small scrap of verbal information can cause a major change in the course of events, as when it indicates that a spouse is being unfaithful or a hostile country has weapons of mass destruction—even if the allegations are untrue. As will be discussed further in Chap. 7, the neuronal configurations with which the brain represents the outside world almost certainly involve some form of symbolic representation in which particular patterns of neuronal connectivity represent particular objects and relationships. However, although intangible ideas and beliefs can cause tangible events that seem to defy the laws of physics, they do not operate entirely separate from them—since information cannot exist without a physical vehicle to convey it. The way our brain's cellular components are arranged and connected represents the information it contains about our thoughts and memories, our feelings and desires, and our understanding of our place in the universe.

Feedback

Cybernetics is the study of the communication and control of feedback in living beings and machines (Wiener 1948). The "magic" that makes these systems work are causal chains of meaningful information—self-correcting feedback signals that regulate certain aspects of an entity's operations. The signals are generated by a discrepancy between a current sensory input and an internally programmed goal, and these are then fed back to turn on or turn off various mechanisms within the entity. As Furman and Gallo (2000, p. 17) note: "An error signal is said to arise from a sensed difference between an *internal reference condition* held for some controlled quality and the current *perceptual condition* as transmitted by one or more sense organs or collections of receptors." Feedback processes enable entities to remain on a specified track by adjusting to changes that occur in their environment. A thermostat, for instance, can keep a room at a constant temperature by switching heating or cooling devices on as needed, pressure receptors in the walls of the major arteries send signals to the brain that regulate blood pressure, and guided missiles alter their course to compensate for sensed deviations from their intended target. Although the transmission of cybernetic signals requires a small amount of energy, the energy that produces the effects they cause is supplied by the recipients, not the signals.[6]

Biological systems are exquisitely self-regulating, for they make extensive use of error-correcting feedback signals to maintain their internal environment within predetermined limits, as well as to regulate their interactions with their external one. Mammalian blood sugar levels, for instance, are controlled by two types of cells in the pancreas, alpha cells that become active when the glucose level gets too low, and beta cells that are activated when it gets too high. The former respond by producing an enzyme that stimulates the liver to release stored glucose and the latter by producing insulin, which helps remove glucose from the circulation. As Jeannerod (1997, p. 4) observes: "Homeostatic systems are closed-loop systems aimed at maintaining the constancy of a fixed inbuilt reference. Their activation depends on the detection of an 'error' between their input level, which monitors the current state of the controlled parameter, and their central level, where the reference value is stored."

The human brain is comprised of a vast array of neuronal feedback circuits that serve to synchronize and control its response to incoming information. Our breathing, heart rate, temperature, muscle action, and metabolism are all closely self-regulated through feedback processes—as also are our interactions with the world around us as we respond to the impact our actions have on it. We are, in fact, caught up in an endless feedback loop in which our behavior is not only shaped by the information we detect, but also helps shape it. We are not merely passive recipients of information, but are also initiators that act on the world to obtain new information about matters of consequence to us. We ask questions, experiment, and try things out in order to shape the informational feedback we receive from the objects and events we encounter.[7]

Goal-Directed Behavior

All living creatures exhibit goal-directed behavior, from the bacteria that seek out food sources to the birds that fly south for the winter, to the stockbrokers who trade on Wall Street. Goal-directed behavior is, in fact, one of the hallmarks of animate entities, something that clearly sets them apart from inanimate ones (Mayr 2004, p. 43). The inability of physical explanations to account for future-oriented behavior has been a major obstacle to understanding the mind as a part of the natural world; for how can future outcomes that have not yet happened cause animals to behave in a certain way? Goal-directed behavior even seems to violate Newton's first law—that a body at rest will remain at rest unless acted upon by some external force—since animals are able to pursue their goals spontaneously, without needing external prompting. In fact, the way they are able to move around under their own steam, rather than just respond to external stimuli, is one of the main ways we distinguish living objects from nonliving ones. It is no wonder that people thought that some sort of mystical process must be involved in causing this, some as yet unidentified vital force or spirit. They were right about there being another process, but not about what it was.

Goal-directed behavior is aimed at bringing an animal's perceptual input into conformity with some type of internal reference (i.e., the goal) through a feedback process. While we set the goals for the devices we design, the goals that shape the behavior of the various species and the individuals within them are established through natural selection and individual experience. The process generally just sets the designated target, not the steps needed to achieve it, especially in animals with complex nervous systems. Our species is unique, however, in that, as well as having instinctive goals that have been shaped by natural selection, we are also able to set a variety of personal ones, like losing weight, saving for retirement, or stopping at the cleaners on the way home (Gazzaniga 2008).

Although the processes of growth and adaptation in animals and plants also appear to be directed at achieving future goals, rather than determined by antecedent states, this is largely an illusion. Biologists have coined the word *teleonomy* to refer to apparently goal-directed, self-regulatory processes in living entities, but the concept is not yet widely accepted—presumably because the idea that effects can precede their causes seems to violate our intuitive understanding of how the world works. However, even though the goals and purposes that characterize teleonomic processes are set in the future, they are represented in the brains and genomes of the involved individuals in the present. Teleonomic effects are thus caused by antecedent informational states, not antecedent physical ones. In the same way that information that accumulates in their genomes over the millennia shapes species-specific goals; information that accumulates in individual brains over a lifetime shapes personal ones.

Shapin (1996) maintains that the mechanical clock became the modern model for understanding the workings of the universe. The way it worked represented a sharp contrast with the animistic view of the world that had been passed down from Aristotle, in which both animate and inanimate objects were endowed with inherent purposes and intentions in order to explain their behavior. The fact that machines could act like intentional agents led people to realize that purposive, goal-directed behavior in humans and other animals could also be explained in terms of mechanical causes. He quotes Robert Boyle's (1627–1691) observation about the famous Strasbourg clock which was completed in 1574: "The several pieces making up that curious engine are so framed and adapted, and are put into such a motion, that though the numerous wheels, and other parts of it, move several ways, and that without any thing either of knowledge or design; yet each part performs its part in order to the various ends, for which it was contrived, as regularly and uniformly as if it knew and were concerned to do its duty." However, as it turns out, the purposive, goal-directed behaviors that characterize living beings, such as building a nest or a house, foraging for food in the wild or the supermarket, or engaging in rituals to find a mate, cannot be explained entirely in physical terms, since they are ultimately caused by information.[8]

Ultimate Causes

As Mayr (1997, p. 118) points out, there are ultimate causes in biology that explain *why* living things function the way they do, not just proximate ones that explain *how* they do it. Proximate causes are the immediately preceding events that trigger a change, while ultimate causes are remote elements in a chain of causality without which the proximate causes would not occur. Physics and chemistry look askance at ultimate causes, because they do not fit into their mechanistic views of causation. The fact that historical causes have no place in the physical sciences is, however, no reason for excluding them from the biologic ones—especially since remote evolutionary and experiential events play a crucial role in determining how living entities function. While both the animate and inanimate worlds obey the laws of physical causation, living organisms are also subject to informational causes that help explain why they act the way they do—why, for example, turtles come ashore to lay their eggs, male deer have large antlers, and bowerbirds build such elaborate nuptial structures. This type of explanation has no counterpart in the more mechanistic approaches of the physical sciences. Natural selection and individual experience shape the information contained in an animal's genome and brain in ways that cannot be explained on a purely physical basis.

Determinism holds that every state of affairs is entirely determined by the state of affairs that immediately precedes it. In theory at least, if we knew everything there was to be known about something, we could predict its future with absolute accuracy, since deterministic systems are entirely predictable once all of the initial conditions and operating rules are specified (Trefil 1997). Some things are usually considered to be unpredictable, however, because their complexity or their inaccessibility makes it impossible to identify enough of their determining factors. The effects caused by information are also unpredictable, at least initially, since there is no direct connection between the information that is detected and the response it generates—for the latter is determined by the recipient, not the initiator. Because the information stored in our brains seems to be too unreachable to ever be completely specified, the causal functions of information may end up being more useful in explaining how our brains work than in predicting how they will function in the future, in much the same way that natural selection explains how living things have evolved, but cannot predict how they will do so in the future.

Most evolutionary changes have been built on conserving previous genomic structures, so that there is a great deal of similarity across the different species in the cellular processes concerned with detecting and responding to information. Many of the complex processes involved in biological informational systems are thus more accessible in simpler organisms—and studies of such creatures have already begun to shed light on the processes involved in more complex ones, including our own. Kandel (2001), for instance, was awarded the Nobel Prize in Physiology or Medicine for 2000 for his work on the molecular causes of learning and memory in the giant sea snail *Aplysia,* which only has about 40,000 neurons (while the human brain has 100 billion of them). Although the focus of this work has so far been on understanding responses to physical stimuli (see Chap. 6), the same principles can be applied to understand how organisms respond to informational ones.

Notes

[1] Salmon (1988), Pearl (2000), and Woodward (2003) review philosophical issues involved in the concept of causation, although they do not address informational causes. Sloman (2005) points out how the notion of cause is used to understand how and why things change.

[2] *Necessary* and *sufficient* are useful concepts for distinguishing between different types of causation. If x is a necessary cause of y, then y will only occur if preceded by x — but the presence of x does not ensure that y will occur, although the presence of y ensures that x must have occurred. Sufficient causes, on the other hand, guarantee the effect, so that if x is a sufficient cause of y, the presence of x guarantees y — but the presence of y does not necessarily mean that x has occurred, since other events may also be able to cause it.

[3] Wilson (1998, p. 266) offers a mathematical theory of information that involves abstract information units called *infons*, but does not explain how these convey meaning or act as causal agents.

[4] Many scientists still believe that vitalism and supernatural forces are the only alternatives to material causation. The eminent biologist E. O. Wilson (1988, p. 266), for instance, maintains that "the central idea of the consilience world view is that all tangible phenomena, from the birth of stars to the workings of social institution, are based on material processes that are ultimately reducible, however long and tortuous the sequences, to the laws of physics."

[5] Schneider and Sagan (2006, p. 21) note: "Traditional physics has not tried to understand things in terms of their ancient causes. The way things worked seldom required an understanding of their history. Mechanical function was understood directly on the basis of present measures."

[6] The word *cybernetics* is derived from the Greek *kybernetes* for helmsman or guide. Gray (2004, p. 304) notes that although "biology in general works by adding to the laws of physics the engineering principles of cybernetics, especially those of feedback and selection by consequences, the principles of cybernetics can't themselves be reduced to the laws of physics—though they must of course respect them." He also observes (p. 35): "One has to consider the passage of nervous impulses around the various complex and interlocking circuits of which the central nervous system is composed as falling simultaneously under the laws of physics and chemistry and under those of cybernetics (or 'information processing', an alternative descriptive language that can be applied to the system level of brain function)."

[7] Powers' (1973) *Perceptual Control Theory* maintains that the aim of animal and human behavior is to control perceptual input in order to minimize error signals between it and internal reference levels. By proposing that an animal's behavior generates the perceptual stimuli it detects, his feedback model reverses the traditional stimulus–response paradigm by emphasizing the way animals initiate behavior.

[8] *Teleos* in Greek means "end" or "goal." Walsh (2008, p. 113) defines teleology as "a mode of explanation in which the presence, occurrence, or nature of some phenomenon is explained by appeal to the goal or end to which it contributes," and reviews the arguments for and against its use in biology. Bigelow et al. (1943) used the term to describe the apparently purposive behavior of animals and machines that were governed by error-correcting feedback processes. Pittendrigh (1958) modified it to *teleonomy* to emphasize that, unlike a number of teleologic goals, these ones are not the result of supernatural processes. Monod (1971, p. 9) proposed that the process was a key feature of biological entities: "Rather than reject this idea (as certain biologists have tried to do) it is indispensable to recognize that it is essential to the very definition of living beings. We shall maintain that the latter are distinct from all other structures or systems present in the universe through this characteristic property, which we shall call teleonomy."

References

Bigelow J, Rosenblueth A, Wiener N (1943) Behavior, purpose and teleology. Philosophy of Science 10: 18–24

Changeux J-P (2004) The Physiology of Truth: Neuroscience and Human Knowledge. Harvard University Press, Cambridge

Dowe P (2000) Physical Causation. Cambridge University Press, Cambridge

Dretske FI (2000) Perception, Knowledge, and Belief: Selected Essays. Cambridge University Press, Cambridge

Furman ME, Gallo F (2000) The Neurophysics of Human Behavior: Explorations at the Interface of the Brain, Mind, Behavior, and Information. CRC Press, Boca Raton

Gazzaniga MS (2008) Human: The Science Behind What Makes Us Unique. HarperCollins, New York

Gray J (2004) Consciousness: Creeping Up on the Hard Problem. Oxford University Press, Oxford

Harold FM (2001) The Way of the Cell: Molecules, Organisms and the Order of Life. Oxford University Press, New York

Haught JF (2006) Is Nature Enough? Meaning and Truth in the Age of Science. Cambridge University Press, Cambridge

Jeannerod M (1997) The Cognitive Neuroscience of Action. Blackwell Publishers, Oxford

Kandel ER (2001) The molecular biology of memory storage: a dialog between genes and synapses. In The Nobel Prizes 2000. Nobel Foundation, Stockholm

Mayr E (1997) This is Biology: The Science of the Living World. Harvard University Press, Cambridge

Mayr E (2004) What Makes Biology Unique? Considerations on the Autonomy of a Scientific Discipline. Cambridge University Press, Cambridge

Monod J (1971) Chance and Necessity. An Essay on the Natural Philosophy of Modern Biology. Alfred P. Knopf, New York

Pearl J (2000) Causality: Models, Reasoning, and Inference. Cambridge University Press, Cambridge

Piaget J, Garcia R (1980) Understanding Causality. Trans. D Miles, M Miles. WW Norton, New York

Pittendrigh CS (1958) Adaptation, natural selection, and behavior. In A Roe, G Simpson (Eds), Behavior and Evolution. Yale University Press, New Haven

Powers WT (1973) Behavior: The Control of Perception. Aldine Publishing, Chicago

Roederer JG (2005) Information and Its Role in Nature. Springer, New York

Rosen R (1991) Life Itself: a Comprehensive Inquiry into the Nature, Origin and Fabrication of Life. Columbia University Press, New York

Salmon WC (1988) Causality and Explanation. Oxford University Press, New York

Schneider ED, Sagan D (2006) Into the Cool: Energy Flow, Thermodynamics and Life. University of Chicago Press, Chicago

Shapin S (1996) The Scientific Revolution. University of Chicago Press, Chicago

Sloman SA (2005) Causal Models: How People Think about the World and Its Alternatives. Oxford University Press, New York

Trefil J (1997) Are We Unique: A Scientist Explores the Unparalleled Intelligence of the Human Mind. John Wiley & Sons, New York

Von Bayer HC (2004) Information: The New Language of Science. Harvard University Press, Cambridge

Walsh D (2008) Teleology. Chap. 5 in M Ruse (Ed), The Oxford Handbook of the Philosophy of Biology. Oxford University Press, New York

Wiener N (1948) Cybernetics: or Control and Communication in the Animal and the Machine. John Wiley & Sons, New York

Wilson EO (1998) Consilience: The Unity of Human Knowledge. Alfred A. Knopf, New York

Woodward J (2003) Making Things Happen: A Theory of Causal Explanation. Oxford University Press, New York

Chapter 4
The Detection of Form

Abstract All living cells and organisms are equipped with specialized receptors that enable them to detect and respond to the form and arrangement of certain types of matter and energy. Each species and cell type has evolved a distinctive set of such receptors that enable it to function successfully within its particular ecological niche. Chemical receptors, like those involved in smell and taste, respond to the molecular patterns of certain substances and are present in virtually every cell and organism in the universe. Physical receptors, like those involved in vision, hearing, and touch, are activated by energy rather than directly by form. Sensory receptors are able to detect and respond to discrete objects, as well as detect the difference between current patterns of sensory input and internal reference ones. Sensation refers to the detection of energy or matter by sensory receptors, perception to the detection of meaningful information in these sensations.

Meaningful information involves the detection of *form* rather than substance (Jablonka 2002, p. 602). These two attributes of nature are inseparable, however, since we cannot perceive form that has no substance, or substance that has no form. Form refers to the way the components of an object or event are arranged, either in space or in time, like the shape of a violin or the sound of a waterfall. It is a function of how the matter and energy that comprise an entity are organized, not a property of the matter and energy themselves. Form is an amorphous *quality* that has no physical effects of its own, since the only way it can have an impact is by being detected by an animate or human-made receptor. This is why physics and chemistry have had a hard time dealing with it—and have relegated it for others to pursue. However, even though philosophers have debated about it since Plato and Aristotle, none of them has recognized its central role in conveying information.[1]

A. Reading, *Meaningful Information: The Bridge Between Biology, Brain, and Behavior*,
SpringerBriefs in Biology 1, DOI 10.1007/978-1-4614-0158-2_4,
© Springer Science+Business Media, LLC 2011

Every living cell and organism is equipped with specialized receptors that enable it to detect and respond to certain forms and arrangements in its surroundings—that is, to information that is meaningful to it. These include surface receptors that detect tactile and chemical information, distance receptors that detect visual and auditory information, and molecular receptors that detect cellular and genetic information. The ability to detect and respond to the form that matter and energy assume is one of the defining characteristics of living entities, something that none of the inanimate ones in nature is able to do. It is the primary means by which animals and plants decode the messages in their genes, coordinate their cellular activities, and adapt to changes in their environment (Gell-Mann 1994, p. 17). Although a profusion of stimuli may impinge on a cell or an organism, only those that trigger a response convey meaningful information (at least as here defined). The response may either be a behavioral one, like fight or flight, a physiological one, like salivating or sweating, or a structural one, like reconfiguring the neural connections involved in learning and memory.

Every species and cell type has evolved a distinctive set of sensory abilities that enables it to function successfully within its particular ecological niche. Natural selection is extremely efficient in the way it makes sure that living things are not burdened by being able to detect information that is not relevant to their needs. As Barkow et al. (1992, p. 9) point out: "A design feature will spread only if it solves adaptive problems better than existing alternatives." Information that is meaningful to one species may thus be meaningless to another. Bats and dolphins, for instance, echolocate, dogs hear ultrasound, bees detect ultraviolet light, fish discern electric fields, and snakes sense infrared radiation, even though most other species lack these abilities. Animals constantly sort through the variety of stimuli that impinge on their sensory receptors, but these only trigger a response when they encounter one of the specific patterns of energy or matter that "turns them on"—that is, when they encounter meaningful information. All of the other stimuli they detect represent meaningless noise as far as they are concerned.[2]

We are, in fact, only able to perceive material objects and events through the detection of form, since our form-detecting senses are the only portals through which information about the outside world enters the brain. We are only able to detect the existence of matter and energy, for instance, by perceiving the alterations they cause in the form of the objects with which they interact. We experience their impact either by the way they *deform* the sensory receptors in our body or by the way they *transform* objects in the world around us. We appreciate their physical properties, like weight, size, and color by the way they *conform* to specified reference standards—and detect motion, growth, and decay by characteristic changes in the form and arrangement of the objects involved. Although all of our physical receptors are able to sense physical stimuli directly, it is the interpretation of the pattern of their sensory input that provides us with meaningful information. We can, for instance, see a flash of light, hear a loud noise, or sense the effects of gravity on a purely physical basis, but have to detect a pattern in the effects they generate to determine whether they contain any meaningful information.

Detecting Information

Nature's devices for detecting form consist of both chemical receptors that respond to the molecular patterns of certain substances, and physical receptors that respond to various types of energy. Chemical receptors, which are present in every cell and organism in the universe, are by far the most common information detectors in nature—and probably the most ancient. They are the mechanism that allows cells to communicate with each other, genetic messages to be translated, and organisms to sense substances in their environment. They are unique in the way they detect and respond to information directly by recognizing form as such, rather than having to recognize the pattern produced by a number of physically activated receptors— which is how the other senses operate. Chemical receptors are comprised of protein molecules that are located both in the membrane that surrounds each cell, where they detect and respond to signals from other cells, and within the cell itself, where they regulate its metabolic activities. Proteins are extremely large molecules that fold on themselves in highly specific, three-dimensional patterns that change when they encounter a small molecule whose complimentary shape fits into their structure. The change that occurs in the protein's spatial configuration when it binds to its particular signaling molecule facilitates (or inhibits) a specific set of chemical reactions in the cell. There are several thousand different protein receptors in a typical cell, each of which is highly selective in that it responds only to a specific chemical substance (called a *ligand*) and activates only a specific cellular response (see Chap. 12). The proteins and their corresponding ligands are like a vast collection of locks and keys that regulate cellular functioning by turning on the physiological activities that are currently needed and turning off those that are not.

Physical receptors differ from chemical ones in that they are initially activated by energy, rather than form. However, it is the *pattern* (i.e., *form*) of the receptors that the physical stimuli activate that represents the information they convey, not the physical stimuli themselves. It is, for example, the pattern of retinal receptors that it activates that enables us to recognize the Mona Lisa, and the pattern of auditory ones that allows us to identify a Beethoven symphony. Physical receptors include external ones that respond to touch (mechanical energy), heat (thermal energy), vision (electromagnetic energy), and sound (pressure wave energy), as well as internal ones that regulate homeostatic functions, such as blood pressure, movement, and body temperature. Physical receptors are not as selective as chemical ones, since they can also be activated by stimuli that do not contain meaningful information—which is the case with a great deal of the sensory input we experience during a typical day.

Although nonmeaningful stimuli cannot trigger an informational response, they can elicit a physical one, like an eye-blink reflex or a reaction to pain. Physical receptors tend to function like radar systems, constantly monitoring environmental stimuli, but only generating an informational response when the pattern a set of them generates matches some type of internal template. Animals need to have a centralized nervous system to be able to detect the meaningful information conveyed by patterns of activated receptors, which is why plants and unicellular organisms only respond to chemical types of information. The information that animals actually perceive is determined by the nature and location of the sensory receptors that are activated, the frequency with which they respond, and the neuronal pathways traveled by the signals they generate (Uttal 2002). Animals derive meaningful information from physical stimuli in two different ways: either by detecting a particular pattern of activated receptors (object detection) or by perceiving similarities or differences between a current activation pattern and a reference one (discrepancy detection).

Object Detection

Object detection depends on recognizing particular patterns of sensory stimulation. The different types of physical receptors employ different strategies for detecting salient objects in their surroundings. For instance, the pattern of tactile receptors that is generated when we touch an object, or are touched by one, provides information about its size, shape, weight, and texture—which enables us to identify and respond to it. The pattern of cochlear receptors in the inner ear that different sound frequencies activate provides similar information about the nature and location of the objects emitting them, just as the pattern of retinal receptors at the back of the eyes enables us to identify visual objects by the contrast between the visible light they emit and the light emitted by their immediate surroundings.[3]

Animals do not have to recognize all the characteristics of a familiar object in order to detect it or identify it as a member of a particular class, since they generally only have to recognize certain key features in order to elicit a response. A herring gull chick, for instance, only has to detect the red spot on its mother's bill to initiate feeding behavior, a newborn goose only has to follow the first moving object it sees in order to become imprinted, and a tick only has to sense the butyric acid present on the skin of a passing mammal in order to drop from its perch.[4] Lorenz (1977) refers to these eliciting signals as Innate Releasing Mechanisms, because of the way they set off fixed motor responses that have adaptive functions for the particular species. The same principles apply to the recognition of learned information patterns, like the way that certain perceptual cues are all we need to detect to be able to identify and respond to familiar people or objects. The process highlights the efficiency with which natural selection operates, for only the minimally necessary amount of information needed to produce the response gets encoded in an animal's recognition templates—as there is no adaptive advantage to encoding a greater amount of detail.

Discrepancy Detection

Discrepancy detection is the other way that meaningful information can be detected. Discrepancy detection indicates that a difference has been detected between a current pattern of sensory input and an internal reference, while conformity detection conveys information that the patterns match each other. The various internal reference patterns include: (a) the immediately preceding sensory input, which provide information about change or constancy, (b) stored memories, which indicate familiarity or unfamiliarity, (c) previously anticipated goals, which lead to satisfaction or disappointment, and (d) stored knowledge structures, which indicate whether the input is likely to be true or useful. In sentient animals, the informational output of these comparisons typically takes the form of a characteristic feeling state, with different types of feelings providing different types of information (see Chap. 14). Feeling states convey evaluative kinds of information, such as whether a perceived object or event is good or bad, right or wrong, strange or familiar, which help shape the individual's response. Feeling states do not provide any information about the stimulus objects themselves, just about whether or not the pattern of neurons they activate is similar to the one referenced. They generate a *qualitative* type of information, which simply indicates how much a detected feature is like or unlike a given appraisal standard. This is the kind of information we receive when we pick up a bag of groceries or a bowling ball—we can tell whether they feel light or heavy, either relative to each other or to an internal standard, but not how much they actually weigh.

We usually define objects and events by the way in which they are like and unlike objects and events we already know and endeavor to understand new ones in terms of ones with which we are already familiar, like when we compare the brain to a computer or call gambling an addiction. We identify *particular* objects and events, however, by the way they differ from the ones that are most like them—that is, by the distinguishing characteristics we overlook when we place them in a category (Model 2003, p. 36). Perceiving conformity can provide a reassuring sense of familiarity with the world, as well as a stable background against which change can be registered. Our sensory systems are tuned to detect the unexpected, so that constant sources of stimulation have diminishing effects as we gradually become habituated to them. What is actually perceived, however, depends on the frame of reference being used. Motion, for instance, is always relative, so that an object can only be perceived to be moving in relation to one that appears stationary. This is why we do not sense how fast the Boeing 747 we are aboard is traveling or feel the earth's movement as it rotates around the sun.

Categories

Detecting conformity between the key aspects of a current stimulus pattern and a reference standard enables us to perceive objects and events as members of a class or category, such as a house, an automobile, or a cocktail party. This enables us to respond in a standardized way to objects and events that have similar features, rather than having to treat each of them as separate entities. The key features that members of a category share function as the meaningful information that generates the response they elicit. The process enables us to respond in an appropriate way when we encounter someone or something with which we are not familiar, like a new teacher, a new concept, or a new appliance. We treat them initially like ones we have previously known—for we have no other way of dealing with them. Identifying people as members of a category, like a sales clerk or a police officer, allows us to interact with them without having to know anything more about them—which is, of course, what we have to do if we want to get to know them personally. We form categories either on a perceptual or a conceptual basis, with the former being based on similarities in appearance and the latter on similarities in function or properties (Sloman and Rips 1998).[5]

The usual way of thinking about perceptual categories is that the neural representations of entities that share common characteristics are linked together as the developing brain becomes more organized (Edelman 2004, p. 49). This may not be correct, however, at least not in all cases, since perceptual categories are initially formed when a number of similar looking objects are perceived to be identical by the developing brain (i.e., they have the same *meaning* to the growing child). Because infants initially recognize objects by just a few key features, they tend to lump all sorts of things together, like calling all pets *doggie* and all round objects *ball*. Their perceptions gradually become more discriminating as various objects attain specific meanings to them, and thus generate specific responses. Increasingly fine discriminations are then made as they grow into adults, depending on the needs and interests of the particular individual. Entomologists, for instance, learn to differentiate between different insects, wine-lovers between different vintages, and concert pianists between different concertos, since these details represent meaningful information to them.

Conceptual categories differ from perceptual ones in that they are based on inferred qualities, rather than perceived ones. They are formed by grouping together items that possess similar properties or functions. These include such everyday categories as clothing, games, and vegetables, all of which are based on similarities in perceived attributes, rather than appearance. Thus, while perceptual categories entail overlooking differences among objects that look alike, conceptual ones involve finding similarities among ones that do not look like each other. A growing child may, for instance, initially refer to a dolphin as a *fish,* based on its appearance, but later learns that it really is a *mammal,* based on its properties and functions. Growing children gradually learn to discriminate between objects that initially seem similar to them— and then respond to each of them differently (Kagan 2002, p. 13).[6]

Sensation Versus Perception

The terms *sensation* and *perception* are often used interchangeably to describe experiences related to our senses of vision, hearing, touch, smell, and taste, as well as internal feelings of pain and discomfort. *Sensation* is used here, however, to refer to the sensory experience generated by a physical stimulus, and *perception* to the information being conveyed by the pattern of the receptors that get activated. Sensation thus refers to the detection of different forms of energy by sensory receptors (e.g., it is green, it croaks, it jumps), and perception to the detection of any meaningful information these sensations convey (e.g., it is a frog). Hearing a siren or an alarm clock go off is a sensation; understanding their meaning is a perception. The meaning of perceived information, in this way of thinking, is indicated simply by the change it produces in the behavior, functioning or neural structure of the perceiver. Perception and sensation are, however, not distinct processes in the chemical senses that regulate cellular and plant activities, since signaling molecules convey information directly, without the need of interpretation.

We do not know how or where sensations are actually turned into perceptions, nor do we have any idea where meaning is detected in the neural pathways that link the various sensory receptors to specific parts of the brain. While sensations are caused by physical stimuli that can be assessed objectively, perceptions are subjective experiences that depend on how individuals interpret the patterns of matter and energy they detect. Perceptions can thus vary from one individual to another, based on their past experiences and beliefs—for how we see the world depends on what our brains make of the patterns our senses detect. The meaning we ascribe to events can thus say more about us at times than it does about the events themselves. This is the basis of the Rorschach test, in which subjects are asked to describe what they see when presented with a standardized set of "inkblot" pictures. Because the way people interpret these ambiguous stimuli is shaped by their personality and emotional predispositions, their responses can be used to assess these traits. The distinction between sensation and perception is also demonstrated in the neurological conditions of *agnosia,* in which affected individuals are able to detect certain sensory stimuli but unable to perceive what they represent. An individual with a visual agnosia, for instance, can copy a drawing of a house or a clock, but cannot tell what they represent, while someone with an auditory agnosia can hear the sound made by a barking dog, but cannot identify what it is.

Notes

[1] Aristotle (384–322 BC) was among the first to distinguish between *matter* and *form.* For him, matter was the undifferentiated primal element, the germinal substance from which all things develop by acquiring a particular form. Pure forms were ideals, essences that existed entirely separate from matter. Medieval scholars saw everything as consisting of both form and matter, with form informing the matter, and matter materializing the form (Borgmann 1999, p. 9). The idea that form can exist separate from matter is still present in our concepts of ethereal spirits and ghosts.

[2] Although we have invented devices like Geiger counters and X-ray machines that can detect types of information that we cannot sense directly, we are still unable to detect some of the information that other animals and plants can. Hughes (2001) outlines some of the special sensory abilities other species possess and explains why studying them is so difficult.

[3] Detecting meaningful information patterns is not always as straightforward as presented here, since they are often embedded in noisy backgrounds from which they have to be differentiated. Recipients tend to make mistakes in doing this when the signal-to-noise ratio is low, either by identifying an information pattern that is not really there or by failing to detect one that is there (Shettleworth 1998, p. 59).

[4] Ticks are blind and find each other by odor. After mating, the female climbs up a tree or bush, where she waits for months or years with sperm and eggs kept in separate storage. When she senses

butyric acid (C_3H_7COOH) wafting up from a mammal below, she releases her grip and falls. If she finds her target, she has a blood meal and then drops to the ground, where the eggs get fertilized and laid in the soil (Sagan and Druyan 1999).

[5]Categorization is also the basis of prejudice and bias, since it enables people to respond to the members of a group as if they were all the same, rather than treating them as individuals. Emotional residues left over from previous relationships can also get transferred to new relationships of a similar category, which is the basis of the transference process that psychoanalysts use to explore their patients' relationships with significant figures in the past.

[6]Kagan (2002) believes that discrepancy detection is one of the main ways that infants gain information about the world. He cites studies that show that the surprise that occurs when they perceive an unexpected discrepancy is even more effective than fear in activating the amygdala region of the brain. Discrepancy detection is essentially the same as the error-detection process in cybernetic systems.

References

Barkow JH, Cosmides L, Tooby J (1992) The Adapted Mind: Evolutionary Psychology and the Generation of Culture. Oxford University Press, New York

Borgmann A (1999) Holding On to Reality: The Nature of Information at the Turn of the Millennium. University of Chicago Press, Chicago

Edelman GM (2004) Wider Than the Sky: The Phenomenal Gift of Consciousness. Yale University Press, New Haven

Gell-Mann M (1994) The Quark and the Jaguar: Adventures in the Simple and the Complex. WH Freeman, New York

Hughes HC (2001) Sensory Exotica: A World Beyond Human Experience. MIT Press, Cambridge

Jablonka E (2002) Information: its interpretation, its inheritance, and its sharing. Philosophy of Science 69: 578–605

Kagan J (2002) Surprise, Uncertainty, and Mental Structures. Harvard University Press, Cambridge

Lorenz K (1977) Behind the Mirror: A Search for a Natural History of Human Knowledge. Trans. R Taylor. Harcourt Brace Jovanovich, New York

Model AH (2003) Imagination and the Meaningful Brain. MIT Press, Cambridge

Sagan C, Druyan A (1999) What thin partitions. In RL Solso (Ed), Mind and Brain Sciences in the 21st Century. MIT Press, Cambridge

Shettleworth SJ (1998) Cognition, Evolution, and Behavior. Oxford University Press, New York

Sloman SA, Rips LJ (1998) Similarity and Symbol in Human Thinking. MIT Press, Cambridge

Uttal WR (2002) A Behaviorist Looks at Form Recognition. Lawrence Erlbum, Malwah

Chapter 5
The Doorways of Perception

Abstract Perception involves sorting out signal patterns that are meaningful from ones that are not, deciphering the information they contain, and then initiating an appropriate response. Chemical information detectors respond directly to form, while the other modalities depend on the pattern of the receptors that detected objects and events activate. Chemical information detection is involved in genetic and cellular regulation, taste and smell sensations, and hormone and pheromone signals. The other sensory modalities respond to various stimuli, like electromagnetic radiation (vision) and air pressure waves (hearing). The different receptors detect different types of meaningful information and respond by transmitting neuronal signals to the brain where it can be interpreted.

Perception is usually defined as the process of acquiring, interpreting, selecting, and organizing sensory information, but it is used here to refer specifically to the process by which an animal detects meaningful information in the signals transmitted from its sensory receptors to its brain. The process involves sorting out signal patterns that are meaningful from ones that are not, deciphering the information they contain, and then initiating an appropriate response, either consciously or otherwise. The neural pulses emitted by activated sensory receptors are not like the electronic patterns that transmit information to our telephones or television sets, since they are simply on/off signals that have no other content. Biological receptors do not transmit the actual information they detect; all they do is send signals to the brain that particular patterns of energy or matter have been detected, and leave it up to the brain to interpret what these mean. Even if we were able to examine these neural signals closely, they would not tell us anything about the objects and events that initiated their activity, for the signals do not contain any information about the identity of the stimulus that triggered them. The only things that differentiate one set of neural signals from another and determine the perception they generate are (a) the location and modality of the receptors that generate them, (b) the pathways along which they travel, (c) the other neurons they activate (or deactivate) along the way,

A. Reading, *Meaningful Information: The Bridge Between Biology, Brain, and Behavior*,   35
SpringerBriefs in Biology 1, DOI 10.1007/978-1-4614-0158-2_5,
© Springer Science+Business Media, LLC 2011

and (d) the locations in the brain where they end up. The neural signals generated by listening to a rock concert or watching a sunset are virtually identical; only the pattern of the receptors that launch them, the neurons with which they interact, and the parts of the brain they activate are different.[1]

Although the process by which information is processed by the sensory receptors is often referred to as *sensory transduction,* no actual transduction takes place—as the detected information pattern is not translated into an equivalent pattern within the nervous system. The information the brain receives is about the state of the sensory nervous system, not the state of the external world. In some miraculous way, it then uses this to reconstruct what is happening in its surroundings. As Mountcastle (1975, p. 109) notes: "Each of us believes himself to live directly within the world that surrounds him, to sense its objects and events precisely, and to live in real and current time. I assert that these are perceptual illusions. Contrarily, each of us confronts the world from a brain linked to what is 'out there' by a few million fragile sensory nerve fibers, our only information channels, our lifelines to reality. Afferent nerve fibres are not high-fidelity recorders, for they accentuate certain stimulus features, neglect others. The central neuron is a story-teller with regard to the nerve fibres, and it is never completely trustworthy."

Most animal species, including our own, make perceptual discriminations over a small but biologically significant range that facilitates their quest for food, safety, sex, and survival. Our sensory receptors detect a great many stimuli that do not convey any meaningful information as far as we are concerned—and thus do not generate a response. Only the objects or events on which we consciously focus our attention are ordinarily able to convey meaningful information. We usually do not perceive any meaningful information in the background details we sense and, as a result, these do not generate a response or get retained in long-term memory. Virtually everything we come to know and understand about ourselves and the world in which we live has entered our brain at one time or another through our various sensory gateways. The reality we come to experience is not captured directly, however, like it is with a camera or a tape recorder, but is instead transmitted from the receptors to the brain simply as patterns of neural impulses that have to be analyzed and interpreted. The brain converts the assortment of sensations we discern into the perceptions we experience in a variety of modality-specific ways, as outlined below.

Taste and Smell

All animals and plants have receptors that can detect chemicals in their immediate environment that may either benefit or harm them. Even single-cell organisms are able to detect and respond differentially to nutrient and toxic substances. Vertebrates have specialized taste and smell receptors that enable them to monitor a variety of chemical compounds in their surroundings, detect the presence of prey or predators, and select foods that are safe to eat. Smell receptors are stimulated by low-weight molecules that get airborne, while taste receptors are activated by larger ones that do not, such as proteins, starches, and fats. Taste and smell require direct contact between a receptor protein and a stimulus molecule with a complementary spatial configuration. When the two bind together, the receptor protein's structure changes in a way that sends a neural impulse to the cortex. The brain then identifies the particular taste or smell by matching the incoming signal against stored templates that have either been encoded genetically or learned from experience (Huang et al. 2006).

Taste sensations arise from populations of chemical receptors located in tiny, bulb-like structures on the surface of the tongue. Five different kinds of taste receptors have been identified in humans: sweet, sour, salty, bitter, and savory, each of which responds to a different chemical modality. Saltiness receptors respond to sodium ions, sourness ones to acids, sweetness ones to sugars, bitterness ones to various alkaloids, and savoriness ones to glutamate. The tongue's taste buds also contain sensory receptors that respond directly to physical stimuli, rather than information, including ones for temperature, pain, irritation, and pressure (Frank 2000). These are believed to be responsible for the spicy, minty, and astringent sensations that some foods elicit, as well as for the appreciation of their texture and consistency. The overall perception of the flavors foods elicit results from the interaction of taste and smell signals, as well as the associations individuals have to them (Taylor and Roberts 2004).[2]

Tastes and smells that animals find pleasant are positively reinforcing and elicit consummatory behaviors, while ones they find unpleasant are aversive and produce avoidance behaviors, even when the subjects are not fully aware of their source. What is experienced as pleasant or unpleasant is, however, not the same for every species. Evolution has shaped the hedonic "value system" of each of them in such a way that objects and events that facilitate their reproductive success and survival come to be associated with pleasant tastes and odors, and ones that hinder these with unpleasant ones. Every species possesses its own portfolio of taste and smell preferences and aversions, since these have been shaped by natural selection. Healthfully ripe fruits taste pleasant to us, for instance, because of the growing levels of sugar they contain, while spoiled meat tastes bad because of increasing levels of acid. Whether a substance smells or tastes good or bad, however, is not a property of the substance, but an attribute applied to it by the particular recipient. The sweet taste of honey or the unpleasant smell of sour milk are just perceptions associated with particular patterns of neuronal activity, not properties of the substances that generate them.

Hormones and Pheromones

Hormones and pheromones are information-conveying substances that are used by animals and plants to coordinate the activities of cells or individuals that are distant from each other. Hormones are chemical signals that one set of cells releases into the circulation or tissues in order for them to be detected by other cells in the same individual, while pheromones are chemical signals that animals and plants release into the environment for them to be detected by other animals and plants. Hormones and pheromones are simply signals whose specificity is determined by the target cells that respond to them, rather than by anything inherent in their own chemical structure. All they do is switch on or off specific routines that are already programmed into their targets and, by doing so, regulate some of their metabolic activities. Hormone and pheromone receptors are formed by various protein molecules, with the composition of each species' receptors determined by its particular genetic makeup. Although closely related species usually have only minor differences is the pheromones they can detect, each one only responds to its own particular set of them. Male moths that fly upwind in response to a sex pheromone released by a female of their own species, for instance, do not respond to the pheromones released by other moth species, even though their chemical structures are closely alike (Wyatt 2003).[3]

There are hundreds of different hormones circulating in the bloodstream and tissues of vertebrate animals. Most of these are secreted by endocrine glands like the thyroid and pancreas, including well-known substances like insulin, epinephrine, and testosterone; but there are many others. The amounts of these chemical messengers in an animal's tissues ebb and flow over time, since they are only released when needed to regulate particular cell functions. The level of thyroid hormone in mammals, for example, is regulated by feedback from the pituitary gland and the hypothalamus, which normally maintains it at the level needed to control the animal's rate of metabolism (Crapo 1985). Many of the hormones in invertebrate animals are related chemically to vertebrate ones, although they do not necessarily have the same functions. Insulin-like substances are present, for instance, in fruit flies, earthworms, protozoa, and fungi, although they are not derived from pancreas cells (Roth et al. 1982). Hormones also play an essential role in coordinating the growth and development of plants, where they regulate cell division, root growth, and shoot development by stimulating or inhibiting the expression of various genes.

Pheromones are hormone-like substances that animals and plants generally use to communicate with other members of their own species. A number of mammals have scent glands that secrete substances that attract potential mates, which is why male deer rub their hind legs against trees and urine from female dogs in heat attracts males from miles away. Female mice that are segregated from males stop having regular sex cycles, but resume them when exposed to the odor of male mouse urine. Insects also secrete pheromones as sexual attractants, territory markers, or alarm substances, and plants emit volatile compounds that can be detected by animals or other plants, including ones that attract pollinators or repel predators (Baldwin et al. 2006). The volatile substances that animals and plants emit can also serve as cues about their location to animals and plants that prey on them, or about the location of ones on which they prey, although these are presumably not the functions for which they were selected. *Dodder*, for instance, is a parasitic plant that locates certain other plants by detecting volatile substances they emit, and then grows toward them, attaches to their stems and leaves, and robs them of needed nutrients (Runyon et al. 2006).

Tactile and Kinesthetic Senses

The human body contains a variety of tactile and kinesthetic receptors that coordinate and stabilize its motor activities. These usually trigger automatic, reflex-like responses that are mediated by spinal cord and subcortical connections, although they can also generate ones that require cortical interpretation, like identifying the texture of an object or detecting an insect crawling on the skin. Kinesthetic senses provide feedback about our spatial orientation and body movement, which enables us to maintain our posture, keep our balance, and move about in the dark. Our muscles, joints, and tendons all contain stretch receptors that generate feedback signals that synchronize our body's actions and coordinate its fine movements, like those needed to thread a needle or catch a baseball.

Our skin contains four types of tactile receptors that combine to provide sensations of heat, cold, light touch, firm pressure, texture, and vibration. They are concentrated in the parts of the body that are most sensitive to touch, such as the lips, mouth, hands, and genitals, each of which is represented in a distinct area in the sensory cortex. The mix of receptors activated by a given stimulus, the pattern of their distribution, the frequency of the signals they generate, and the locations in the cortex to which they send their signals, all provide information about its nature and identity. Pain sensations, however, are not generated by tactile receptors, but by free nerve endings that respond to substances that damaged tissues release. Mammals also have pressure receptors in the walls of their major arteries that respond to changes in blood pressure by altering the frequency of the impulses they transmit to the brain, which then modifies the sympathetic and parasympathetic signals that regulate heart rate and blood vessel diameter (Ádám 1998).

Hearing

Hearing is essentially a mechanical process, although the perception of meaningful information involves cortical interpretation of the signals the auditory receptors generate. Sound is produced by pressure waves transmitted through air or water to the inner ear, where they cause movement of the hair cells in the cochlea. Each group of hair cells responds best to a particular wave frequency and transmits a distinctive pattern of neuronal activity to the cortex. The amplitude of sound waves determines their loudness, the frequency their pitch, and the shape their quality or timbre. We are ordinarily able to hear sounds with frequencies between about 20 and 20,000 vibrations per second, but several species can detect much higher ones. Dolphins and bats, for instance, use ultrasound waves to locate objects by echo-detection.

Although our auditory receptors react physically to the sound waves that impinge on them, only certain patterns of activation contain meaningful information for the particular recipients, with the rest being merely noise. The cortex sorts and interprets the various signals it receives, including those conveyed by spoken language, which is in a code that first has to be deciphered. Sound is the main vehicle that many birds and mammals use to communicate warnings, indicate territory, and make mating calls to other members of their own species. There is no sound in a vacuum, however, since there are no molecules to vibrate and bump into each other. Strictly speaking, there is also no sound when a tree falls in the forest and no one is there to hear it, since sound is a perceptual experience, not a physical one.

Vision

The retina at the back of the vertebrate eye contains rod and cone cells that can detect the wavelength of light reflected off the surface of external objects. When light waves fall on these receptors, the special pigments they contain undergo a chemical change that triggers the transmission of a neural impulse to the visual cortex at the rear of the brain.[4] The human eye contains about 6 million cone cells that can detect color and fine detail, as well as about 125 million rod cells that are sensitive to movement and low-intensity light. Most of the cones are located in the fovea, a central part of the retina that is used for fine vision, while the rods are dispersed throughout the entire retina. Visual perception is an extremely complicated process, for not only do we have to reconstruct a perceived image from the various features we detect, we also have to determine whether it contains any information that is meaningful to us. We do this by searching for salient patterns and discrepancies between the current image and reference ones. This pattern-matching process allows us to identify particular objects by their distinctive shape and composition, even when the size of their retinal images varies. We are also able to see images in our "mind's eye" in the absence of external stimulation by creating them from stored memories and representations. Scenes that are imagined or recalled appear to activate the same brain regions that comparable perceptual images set off, and the perception of another person doing something appears to involve the same motor neurons as doing it oneself (Gallese et al. 1996).[5]

Light entering the eye is focused onto the retina by the lens, but the pattern of retinal receptors that are activated is not the same as the one perceived by the brain. Several factors contribute to this discrepancy: (a) the image perceived by the brain does not include the blinks, eye movements, and saccades that interrupt the pattern of retinal activation, (b) the "blind spot" on the retina where the optic nerve exits the eyeball is not evident in the perceived image, even though it has no receptors, (c) the network of blood vessels and nerve cells that lies immediately in front of the retinal receptors is not apparent in the perceived image, even though it interferes with the light reaching the receptors, and (d) the perceived image is right side up and three-dimensional, while the retinal one is upside down and two-dimensional. The primate visual system is remarkable in the way it can accurately identify given objects at different distances, in different illuminations, in different contexts, at different orientations, and with different colorations, even though the actual pattern of light falling on the retina is different in each of these situations (Kosslyn 1994).

The retinal image is not transmitted directly to the brain, but is deconstructed into its component features, which are then analyzed for meaningful patterns as they are re-integrated in the cortex (Blake and Sekuler 2005). The visual cortex contains specialized feature-detecting cells that respond to particular orientations, contours, edges, and directions of movement, as well as to various shapes, like hands and faces (Galaburda et al. 2002). Some cortical cells respond best to light lines on dark backgrounds, some to dark lines on light backgrounds, some to horizontal rectangles, and others to vertical ones (Hubel and Wiesel 2005). This arrangement allows us to quickly perceive rough categorizations of objects, like recognizing that an object is an animal, but additional processing is required to identify what kind of animal it is, and even more to recognize it as a specific individual (Brincat and Connor 2006).

Space and Time

Most animals have some type of awareness of their location in space and time, even though they do not appear to have sensory receptors that can detect this information directly. Migratory birds are able to travel thousands of miles to return to the same breeding site every year, salmon can find their way back to their birth river to spawn, and penguins return directly to their own nest after foraging at sea for several days, even though it lies in the midst of a hundred thousand seemingly identical ones. Although we marvel at these feats, we know little about how they are accomplished. Some animals appear to derive information about space and time by combining data from internal clocks with knowledge of the position of the sun, the stars, or the earth's magnetic axis, but how they do it is not known.

Hunting, foraging, and migrating animals need to have an effective means of finding their way home after they have traveled to distant locations. While some of them locate their way back over short distances by noticing landmarks or leaving a chemical trail, these are not practical solutions for the relatively long distances that others travel. The desert ant *Anaglyphic bicolor*, for instance, searches for food on an erratic-looking path that may take it as far as 100 m from its nest, but turns and heads directly back to the nest once it finds what it is looking for. These tiny creatures apparently create some type of internal map that keeps track of their location relative to their nest by computing the distance they have traveled and the changes in their direction relative to the sun. When they are artificially displaced some distance from where they have just found food, they turn and head in the direction that would have taken them home from where they were before they were moved, rather than in the direction that takes them home from where they have been relocated (Wehner and Srinivasan 1981).

Some species seem able to estimate the time of day from internal clocks or the position of the sun, while others appear to assess elapsed time periods from a sense of their own rate of activity. Anything that changes in a highly uniform manner can be used to measure time, like the daily cycle of the sun or the oscillations of a particular neural circuit. The activity cycles of nocturnal animals, like hamsters and cockroaches, still show a recurrent period of approximately 24 h when they are deprived of light cues. These cycles gradually diverge from regular clock time, however, when the animals are kept in the dark for lengthy periods, although they can be readily reset by subsequent exposure to light. We have similar circadian rhythms that determine the timing of a number of our physiological functions, as well as cause the jet lag we experience when we travel across time zones. As James (1890) points out, we do not experience time directly, but infer it from our experience of events occurring in succession with an interval of varying length between them. Our subjective sense of the passage of time is based on the rate of change we perceive in our sensory input. Some events seem to take longer than clock time, while others seem to take less, since time seems to pass faster when filled with interesting or exciting experiences, and to drag when devoid of them.

Notes

[1] Descriptions of the physiological mechanisms involved in sensation and perception can be found in Brown and Deffenbacher (1979), Smith (2000), Goldstein (2002), Coren et al. (2003), Wolfe et al. (2005), and Blake and Sekuler (2005).

[2] There are about a thousand different types of odor receptors in the human nose, with several thousand copies of each being dispersed throughout the nasal epithelium (Dulac 2006). We can recognize about 10,000 different odors from the combination of receptors they activate. According to Buck (2005), the neurons in the olfactory cortex seem to function as analog detectors that match the input patterns from the olfactory receptors against inbuilt templates.

[3] It is not clear whether pheromones play a significant role in humans (Wysocki and Preti 2009). However, the menstrual cycles of female coeds tend to become synchronized after 4–7 months of living together, and seemingly odorless compounds from the armpits of women have been shown to synchronize the menstrual cycles of other women (McCarthy and Becker 2002).

[4] Each cone cell contains one of three slightly different pigments that responds best to either red, blue, or green light and together provide the range of colors we perceive. Visible light is an electromagnetic radiation with wavelengths between 4,100Å (Ångstrom) for violet and 6,800 Å for red, so that a rose that reflects light with a wavelength near the upper end of this spectrum is perceived as red, while a leaf that reflects light near the middle is perceived as green. Ultraviolet and X-rays have shorter wavelengths, while infrared, microwaves, and radio have longer ones, but these do not activate our visual receptors. A number of invertebrates have light-sensitive cells that help orient them towards the sun.

[5] *Mirror neuron* is a term that has been applied to certain motor neurons in primates that have been found to fire both when the animals execute a goal-related action, like grasping an object, and when they observe other animals performing the same task. While emphasis has been placed on the role of these cells in facilitating imitative behavior and empathy, their actual function remains unclear. It is possible that they merely indicate the way the brain "multitasks" by using the same neuronal configurations to represent the same actions, no matter whether they are being performed, imagined, or observed.

References

Ádám G (1998) Visceral Perception: Understanding Internal Cognition. Plenum, New York

Baldwin IT, Halitschke R, Paschold A, von Dahl CC, Preston CA (2006) Volatile signaling in plant-plant interactions: "talking trees" in the genomics era. Science 311: 812–815

Blake R, Sekuler R (2005) Perception, 5th edn. McGraw-Hill, New York

Brincat SL, Connor CE (2006) Dynamic shape synthesis in posterior inferotemporal cortex. Neuron 49: 17–24

Brown E, Deffenbacher K (1979) Perception and the Senses. Oxford University Press, New York

Buck LB (2005) Unraveling the sense of smell. In T Frängsmyr (Ed), The Nobel Prizes 2004. Nobel Foundation, Stockholm

Coren S, Ward LM, Enns JT (2003) Sensation & Perception, 6th edn. John Wiley & Sons, New York

Crapo L (1985) Hormones: The Messengers of Life. WH Freeman, New York

Dulac C (2006) Charting olfactory maps. Science 314: 606–07

Frank RA (2000) Neurobiology: Tepid tastes, Nature 403: 837–839

Galaburda AM, Kosslyn SM, Christen Y (2002) The Languages of the Brain. Harvard University Press, Cambridge

Gallese V, Fadiga L, Fogassi I, Rizzolatti G (1996) Action recognition in the premotor cortex. Brain 119: 593–609

Goldstein EB (2002) Sensation and Perception, 6th edn. Wadsworth Publishing, Pacific Grove

Huang AL et al. (2006) The cells and logic for mammalian sour taste detection. Nature 442: 934–938

Hubel DH, Wiesel TN (2005) Brain and Visual Perception. Oxford University Press, New York

James W (1890) Principles of Psychology. Henry Holt, New York

Kosslyn SM (1994) Images and the Brain: The Resolution of the Imagery Debate. MIT Press, Cambridge

McCarthy MM, Becker JB (2002) Neuroendocrinology of sexual behavior in the female. In JB Becker, SM Breedlove, D Crews, MM McCarthy (Eds), Behavioral Endocrinology, 2nd edn. MIT Press, Cambridge

Mountcastle VB (1975) The view from within: Pathways to the study of perception. Johns Hopkins Med J 136: 109–131

Roth J, LeRoith D, Shiloach J et al. (1982) The evolutionary origins of hormones, neurotransmitters and other extracellular chemical messengers. NEJM 306: 523–527

Runyon JB, Mescher MC, DeMorales CM (2006) Volatile chemical cues guide host location and host selection by parasitic plants. Science 313: 1964–1967

Smith CUM (2000) Biology of Sensory Systems. John Wiley & Sons, New York

Taylor AJ, Roberts DD (2004) Flavor Perception. Blackwell Publishing, Oxford

Wehner R, Srinivasan MV (1981) Searching behavior of desert ants, genus Cataglyphis (Formicidae, Hymenoptera). J Comp Physiol 142: 315–338

Wolfe JM, Kluender KR, Levi DM, Bartoshuket LM (2005) Sensation and Perception. Sinauer Associates, Sunderland

Wyatt TD (2003) Pheromones and Animal Behaviour: Communication by Smell and Taste. Cambridge University Press, Cambridge

Wysocki CJ, Preti G (2009) Human Pheromones: What's Purported, What's Supported. The Sense of Smell Institute White Paper, http://senseofsmell.org

Chapter 6
Response Systems

Abstract Biological detection and response units have coevolved, since detecting information offers no adaptive advantage on its own. Fixed responses to information in a given species are usually referred to as instincts because they involve "hard-wired" neuronal pathways that have developed through natural selection. Learned and remembered information modify the neuronal circuits of more developed species, so that their responses to specific information patterns become varied. Meaning is an attribute supplied by the perceiver, not an inherent property of perceived objects or events. It is a function of the response they generate, and can be inferred on this basis. Enhanced ways of detecting and responding to information are the primary ways living entities achieve adaptive superiority, either at the species or the individual level.

Meaningful information is defined by its ability to generate a response in the recipient—either a behavioral one, like approach or avoidance, a physiological one, like the release of a hormone, or a structural one, like a change in neuronal connectivity. Any sensory input that fails to do this is considered to be meaningless as far as the particular individual is concerned. The ability to respond to meaningful information is part of the circuitry that enables living organisms to maintain their internal homeostasis, regulate their behavior, and adapt to changes in their environment. Successful adaptation depends on linking perceptions to beneficial responses, since these connections are what make it possible for organisms to locate food, find mates, and avoid danger (Noë 2004). The biological systems for detecting and responding to information have almost certainly coevolved, for there is no selective advantage in being able to detect or respond to information that is of no consequence to the recipient (Berthoz 2000, p. 3).[1]

A. Reading, *Meaningful Information: The Bridge Between Biology, Brain, and Behavior*, SpringerBriefs in Biology 1, DOI 10.1007/978-1-4614-0158-2_6, © Springer Science+Business Media, LLC 2011

Circuits

Meaningful information is the input arm of detection–response circuits that enable organisms to adapt to changes in their internal and external environments. These sensory–motor linkages are presumably the units of evolutionary change, since they are an essential feature of biologically adaptive systems. Tweed (2003, p. 4) believes they are the key to understanding the brain, which he conceives as composed of a hierarchically arranged set of them: "All the brain's inner workings, all its thoughts and feelings, membranes and synapses, neurotransmitters and receptors, are there only to achieve the right ratio between sensory input and motor output. It was by weighing sensorimotor transformations one against the other that natural selection made us what we are." Gallistel (1980, p. 392) also envisions the brain as consisting of a series of interacting sensory–motor circuits, together with internal "oscillators" that can generate neural activity independent of any current sensory input, so that animals can also initiate actions on their own.

Detection–response circuits in unicellular organisms automatically trigger predetermined routines when they encounter substances in their immediate surroundings that are "meaningful" to them. *Escherichia coli*, for example, is a single-cell bacterium that typically resides in the intestine, where it swims around to locate nutrients and avoid harm. Each of these microorganisms has a number of specialized chemical receptors in the membrane that encloses it, some of which detect the sugars it needs for energy, some the amino acids it requires to make proteins, and some the toxic chemicals that can injure it. When these receptors detect their matching substance, they initiate an internal chain of biochemical reactions that alters the rotation of the organism's propeller-like flagellae, thereby causing it either to swim toward a beneficial substance or tumble away from a harmful one (Adler 1966). The relatively simple circuits that regulate these unicellular organisms likely represent the building blocks of the complex information processing systems that characterize multicellular animals. Almost all multicellular animals have chains of specialized information-relaying cells (i.e., neurons) interposed between their sensory receptors and their response systems, which allow them to modify how they respond, based on their past experience and associations.

Innate Responses

When the response to a particular information pattern is relatively fixed in a species, it is usually referred to as an instinct. These innate responses can involve complex patterns of behavior, even though simple on/off signals are all that is needed to set them off. The nature of these responses cannot be predicted from the nature of the signals, at least not initially, since it is determined by the recipients. There is, for instance, no a priori reason for predicting that the red spot on the herring gull's bill would elicit a feeding response in her chicks or that the color of a male stickleback's belly would induce a female to mate. Even though they have been shaped by evolution, innate behaviors may require environmental cues to function properly. A mother rat, for instance, normally builds a nest before bearing offspring and then grooms her newborn pups, even when she has been raised in isolation from other female rats. She will not build a nest, however, if raised in a cage that has no materials to carry in her mouth, and will not groom her young if reared with a collar that prevents her from licking herself (Cziko 2000, p. 128).

Innate behavior generally involves "hard-wired" neuronal connections between the involved detection and response units. Frogs, for example, contain specialized ganglion cells in their retina that are specifically tuned to detect small, dark, quick-moving objects in their field of vision. When these "bug-detector" cells are activated, they send neural signals to the motor system that cause the frog to turn toward the object and strike at its precise location. Because these specialized cells are unable to respond to stationary objects, frogs that are surrounded by foods that do not move can starve to death (Lettvin et al. 1959). The triggering of such fixed action patterns by innate releasing mechanisms is widespread throughout the animal kingdom, where it plays a role in a number of key behaviors, including imprinting, courtship rituals, and predator avoidance. Innate responses ordinarily occur automatically and in exactly the same manner in every member of a particular species (and often particular gender), like hatching sea turtles heading toward the ocean, nest-building in birds, and mothering behavior in mammals.

Learned Responses

As organisms increase in complexity, they develop centralized nervous systems that allow them to vary how they respond to meaningful information by including input from stored memories, linked associations, internal models, and contextual factors. Learned and remembered information increasingly come to modify the detection–response circuits of the more highly developed species, so that the way different members respond to given informational patterns becomes more diverse over time. Learning essentially involves rearranging the patterns of neural connectivity between receptors and effectors, so that an organism no longer responds the same way to the same information. Previously meaningless patterns of matter and energy can become meaningful and generate a response as the result of learning, and previously meaningful ones can have their meaning changed, so that they generate a different response. The response to learned information is usually not completely fixed, however, so that a learned goal may be set, but not the steps needed to achieve it. Rats, for instance, do not take exactly the same path every time they run through a maze they have learned, just as shoppers do not go over exactly the same route every time they drive to the supermarket (Cziko 2000, p. 44). Being able to modify how they respond to detected information endows individuals with a useful degree of flexibility, especially in responding to circumstances that were not present in the environments that shaped their current genetic composition.[2]

Learning can also modify innate circuits, as shown by the gill-withdrawal reflex of the marine mollusk *Aplysia*. Habituation (decreased responsiveness after repeated stimulation) weakens the in-built connections between the sensory and motor neurons that control this reflex, while sensitization (stimulating neighboring sensory areas) strengthens them. During short-term learning, the transmission of impulses over the neuronal pathway that links the stimulus to the response is temporarily altered; but, for more permanent learning to take place, new synaptic connections have to be formed (Kandel 2006). Although this particular reflex involves a physical stimulus, similar neuronal changes are apparently involved in the response to informational patterns, including those in more highly developed species.

Most invertebrate species are able to modify their preprogrammed circuits as a result of experience. Bees, for instance, can learn the location of a variety of food sources and then tell their hive-mates about their distance and direction by modifying elements of their built-in dance language. The types of information that are meaningful to invertebrates seem to be limited to finding food, locating mates, and avoiding harm, a prioritization presumably necessitated by the size of their nervous systems (Prete 2004). Although some plants can respond to mechanical stimuli, like a Venus flytrap catching an insect or a sunflower turning toward the sun, they are unable to move or change what they are doing in response to informational cues. They have little need to be mobile, however, since they get their energy from the sun without having to go searching for substances that can supply it. Plants also have no capacity for memory or learning, since they do not have any neurons to modify how they respond. The only information patterns they are able to detect are the chemical signals involved in cellular functions, but they are all they need to grow and prosper.

Meaning

Meaning is itself a word that has several meanings, including being a synonym for (a) *purpose*, as in "what is the meaning of life," (b) *significance,* as in "what is the meaning of these findings," and (c) *definition,* as in "what is the meaning of *oxymoron*" (Millikan 2006). The term is used here, however, to refer solely to the meaning of the effect an object or event elicits in a cell or organism. Meaning, in this sense, is an inferred attribute of the response that patterns of matter and energy generate in a recipient. This usage is essentially free of the usual linguistic and metaphysical connotations of "meaning," since it does not have to be consciously perceived and can be applied across the entire spectrum of information-detecting entities, from cells and microorganisms to plants, animals, and intelligent machines (Menant 2003). Defining something in terms of the effect it produces follows the pragmatic approach recommended by Peirce (1878, p. 289), who advised: "Consider what effects that might conceivably have practical bearings. To develop its meaning, we have, therefore, simply to determine what habits it produces, for what a thing means is simply what habits it involves."[3]

The meaning of perceived information in this way of thinking is a function of the effect it has on the recipients, so that detected stimuli that have no effect are meaningless as far as they are concerned. Meaning is not an inherent property of the perceived objects or events; it is an inference that denotes the effects they have on particular perceivers, which is why it is always personal (Ogden and Richards 1923). The meaning a detected object or event has for given individuals can thus be defined by the response it generates in them, and can be inferred by outside observers on this basis. The meaning to *E. coli* of certain sugars in its environment is thus something to swim toward, the meaning to a frog of a small dark object moving in front of it is something to catch and eat, and the meaning of a stop sign or red traffic light is something to get us to bring our vehicle to a halt. Since the effect that meaningful information has on particular recipients is determined by how each of them sees and understands it, their responses cannot be explained or predicted by objective means.[4]

Effects that are caused by mechanical forces, like the workings of an engine, the orbits of the planets, or the throw of a dice are usually not associated with meaning (although mystics and astrologers still thinks they are). Meaning, in the sense used here, applies only to information-caused events. While Newton's physical response to the proverbial apple falling on his head was based on mechanical principles, his informational one was determined by the meaning he gave to the event. The meaning of a $10 bill, a flag at half-staff, or a police car's siren exists in the mind of the beholders, not in the objects themselves. Perceived information does not even have to be true to be meaningful, which is why the misinformation that spin-doctors, talk shows, and advertisements often disseminate can have a significant effect on people, as long as they believe it is true.[5]

A New Model

The traditional stimulus–response (S→R) model that characterized psychology for many years may need to be modified, since behavioral responses are generated by meaningful information, not stimuli. A meaningful information–response (I→R) model more accurately portrays that it is the detection of form, not energy, that generates the response. It also makes it easier to conceptualize how intangible mental events, like thoughts and beliefs, can cause tangible behavioral effects, since it allows stored neuronal patterns to generate responses. Although physical stimuli can trigger reflex responses, meaningful information is needed to generate behavioral ones. It seems likely, however, that I→R circuits evolved from S→R ones, since the less highly developed species respond to chemical information in a reflex-like manner. Pavlov's (1927) belief that instincts are nothing more than complex reflexes with a longer chain of connections may prove to be correct, especially if the chains involve informational patterns rather than the physical stimuli he envisioned.

The I→R model also offers an alternate perspective on learning. In classical conditioning, repeatedly pairing an informational pattern that normally evokes a certain response with a neutral one that does not eventually causes the neutral pattern to elicit the response by itself. Pairing the presentation of meat with the sound of a bell, for instance, results after a short while in the bell alone being able to make the dog salivate. In this type of learning, information that was previously meaningless (e.g., the sound of the bell) comes to acquire meaning (i.e., produce a response) by being linked to a sensory input that is already meaningful to an animal. In operant conditioning, on the other hand, the frequency of a particular behavior can be increased by following it with a pleasurable physical or informational event (a reward) or reduced by following it with an unpleasant one (punishment). Both types of learning involve linking previously unrelated circuits in the nervous system, so that the information conveyed by an object or event changes its meaning to the individual, and thus generates a different response.

Selective Advantages

The engine that drives biological evolution is the differential survival of the genes of individuals who are better able to meet challenges in their physical and social environments (Bowler 2009). Enhanced information-processing abilities are the primary way that living entities achieve adaptive and reproductive superiority, both at the individual and the species level. Animals and plants with more effective ways of detecting and responding to information have an adaptive advantage over less-capable individuals, especially when resources are limited. A great number of the differences between the various species are thus related to the sensory and motor systems they have developed. The journey from the first primordial life forms to the present day has essentially been shaped by the emergence and selection of increasingly effective methods of acquiring essential resources and avoiding potential dangers. As Carroll (2005, p. 168) observes: "the drama of evolution has been a race to find better, faster, lighter, stronger or more nimble limbs with which to swim, walk, run, hop, breathe, burrow, or fly, or to grab, crush, swallow, poke, filter, suck or chew food." It has also been a race to find more effective, more sensitive, and more relevant ways of detecting information that can help organisms obtain food and other resources, avoid danger, and compete for opportunities to reproduce.

Living things are different from nonliving ones. They *do* things, like reproduce themselves, import free energy from their surroundings, and respond to what they encounter in ways that promote their own survival and genetic perpetuation (Schrödinger 1946). Most of the evolutionary changes that have taken place since the first living things crawled out of the proverbial swamp have been associated with the development of more effective ways of detecting and responding to information, for information-processing skills are the currency of biological adaptation and survival. This is why so many of the changes that natural selection has brought about have been in the development of enhanced ways of perceiving and utilizing meaningful information. Improved sensory and motor skills do not provide any selective advantage on their own, however, since evolving organisms also have to have a way of determining whether what they sense is good or bad for them, and some way of basing their responses on these assessments. Individuals who were better at this were more successful at passing their genes (and related skills) on to succeeding generations. As Glimcher (2003, p. 172) notes: "At a very fundamental level the goal of all behavior must be to use sensory data and stored knowledge of the structure of the world to produce motor responses that are adaptive."[6]

Notes

[1] Because the changes in the nervous system that accompany memory and learning are not directly observable, it is not always possible to tell whether a given information pattern has triggered a neuronal response in individuals who detect it, that is, whether it represents meaningful information to them. Although a great deal of what is called "information" in today's world seems to have no lasting effect on most people, it is always possible it has affected them in ways that are not readily apparent. We have no way of knowing, for instance, whether some of the things that people see and hear on television or read in the newspaper produce subtle changes in their patterns of neuronal connectivity that cause them to be different in ways that they would not have otherwise.

[2] There may be an adaptive advantage in leaving the steps needed to achieve a particular goal up to chance. Domenici et al. (2008), for instance, report that the direction cockroaches use to escape from a predator is essentially unpredictable, which prevents the predator from learning the path they will take.

[3] Most of the philosophical discussion of *meaning* is concerned with the meaning of language-related concepts, not information-related ones (Richards 2003). The fields of Semiotics and Semantics (studies of signs, symbols, and language) examine how the meanings of communicated messages are constructed and understood, some of which is discussed in Chap. 11.

[4] We are the only species that searches for a meaning to our lives, since, as Frankl (1984) notes, the feeling that everything is meaningless leads to an existential type of despair. We need to see our lives as having a meaning and a purpose, and are uncomfortable thinking they may not. This is part of the price we pay for being able to infer meanings and purposes in the things we encounter, since it seems unlikely that other species worry about these existential concerns.

[5] Dretske (1981) maintains that falsehoods are not information, since he believes that information has to be capable of yielding knowledge to the receiver, which he defines as justified true belief. The trouble with this definition is that everyone believes that their own beliefs are justified, so that different individuals believe that different things are true.

[6] Life on earth began about 3.5 billion years ago as tiny, single-celled, marine organisms that were scattered randomly about by the currents and tides, much like present day plankton, without any way of controlling where they were going or what they were doing. It took almost another 2 billion years for organisms that could move under their own steam to evolve. These were unicellular microbes that could move along chemical gradients to find nutrients or avoid toxins (chemotaxis), or along light gradients to facilitate photosynthesis (phototaxis), but were not able to otherwise control what they were doing. The multicellular organisms that emerged about a million years later heralded the development of specialized sensory and motor cells. These new skills greatly enhanced the ability of these tiny creatures to detect and respond to information about objects and events that could help or harm them. Those that were more successful at these tasks proliferated over the next half billion years in ways that lead to the emergence of the first vertebrates. These were free-swimming aquatic creatures with a variety of sensory receptors that sent signals to a central brain which evaluated the information they contained and signaled the animal's muscles and glands to respond accordingly. As species with more complex sensory and motor skills continued to evolve, terrestrial mammals appeared about 200 million years ago, and our first hominid ancestors about 2.5 million years ago.

References

Adler J (1966) Chemotaxis in bacteria. Science 153: 708–716

Berthoz A (2000) The Brain's Sense of Movement. Trans G. Weiss. Harvard University Press, Cambridge

Bowler PJ (2009) Darwin's originality. Science 323: 223–226

Carroll SB (2005) Endless Forms Most Beautiful: The New Science of Evo Devo and the Making of the Animal Kingdom. WW Norton, New York

Cziko G (2000) The Things We Do: Using the Lessons of Bernard and Darwin to Understand the What, How, and Why of Our Behavior. MIT Press, Cambridge

Domenici P, Booth, D, Blagburn, JM, Bacon JP (2008) Cockroaches keep predators guessing by using preferred escape trajectories. Current Biology 18: 792–796

Dretske FI (1981) Knowledge and the Flow of Information. MIT Press, Cambridge

Frankl VE (1984) Man's Search for Meaning. Washington Square Press, New York

Gallistel CR (1980) The Organization of Action: A New Synthesis. Lawrence Erlbaum Associates, Hillsdale

Glimcher PW (2003) Decisions, Uncertainty, and the Brain: The Science of Neuroeconomics. MIT Press, Cambridge

Kandel ER (2006) In Search of Memory: the Emergence of a Science of Mind. W.W. Norton, New York

Lettvin JY, Maturana RR, McCulloch WS, Pitts WH (1959) What the frog's eye tells the frog's brain. Proc. Inst. Rad. Eng. 47: 1940–51 (Reprinted in McCulloch WS 1965, Embodiments of the Mind. MIT Press, Cambridge)

Menant C (2003) Information and meaning, Entropy 5: 193–204

Millikan RG (2006) Varieties of Meaning. MIT Press, Cambridge

Noë A (2004) Action in Perception. MIT Press, Cambridge

Ogden CK, Richards IA (1923, 1989) The Meaning of Meaning: A Study of the Influence of Language upon Thought and of the Science of Symbolism. Mariner Books, New York

Pavlov IP (1927) Conditioned Reflexes: An Investigation of the Physiological Activity of the Cerebral Cortex. Trans. G. V. Anrep. Oxford University Press, London

Peirce CS (1878) How to make our ideas clear. Popular Science Monthly 12: 286–302

Prete F (2004) Complex Worlds from Simpler Nervous Systems. MIT Press, Cambridge

Richards M (2003) Meaning. Blackwell, Malden

Schrödinger E (1946) What is Life? Macmillan, London

Tweed D (2003) Microcosms of the Brain: What Sensorimotor Systems Reveal About the Mind. Oxford University Press, Oxford

Chapter 7
Cognitive Processing

Abstract Cognition consists of the mental processes that mediate between the detection of meaningful information and the response it generates. It includes a range of information-processing activities, including thinking, reasoning, and decision-making. One of the things that is usually left out of cognitive models of information processing is, however, a clear definition of what exactly is being processed. Analog processing involves looking for similarities and differences (i.e., analogies) between various information sets, while digital processing involves the manipulation of symbols according to predetermined rules. Vision and hearing process both digital and analog information, while taste, smell, touch, and kinesthesia depend on analog information. Symbols are discrete patterns of energy or matter that uniquely represent some other entity, convey the same information and meaning it does, and generate the same response. No one knows how the brain creates the symbols it uses to portray concrete objects or represent their properties and relationships.

Cognition is comprised of the mental processes that mediate between the detection of meaningful information and the response it generates. It entails a wide range of information-processing activities in humans, including thinking, attending, reasoning, decision-making, problem-solving, planning, reminiscing, and imagining, although it does not usually include emotional reactions or involuntary responses. It represents the *throughput* functions that take place in the cerebral cortex and determine how individuals analyze and interpret the information they detect, both consciously and otherwise. Because they are more accessible in our own species, this chapter focuses on the processes that underlie human cognitive abilities.[1]

One of the concerns that is usually omitted from the current information-processing models of cognition is a clear definition of what exactly it is that is being processed. Most people seem to assume that the nervous system transmits the same information their sensory receptors detect, and that the brain processes this much the same way computers do. Meaningful information, however, is not transmitted directly from the sensory receptors to the brain, but is instead transformed into

A. Reading, *Meaningful Information: The Bridge Between Biology, Brain, and Behavior*,   53
SpringerBriefs in Biology 1, DOI 10.1007/978-1-4614-0158-2_7,
© Springer Science+Business Media, LLC 2011

patterns of on/off neural signals whose meaning has to be interpreted by the recipient. As discussed in Chap. 4, our nervous system recognizes meaningful information either by detecting a particular pattern of activated receptors or by detecting discrepancies between a current pattern and a reference one. These two ways of perceiving meaningful information enable us to process both discrete (digital) and continuous (analog) patterns of matter and energy.

Digital and Analog

The human brain has two basic ways of representing and processing information, which are referred to here as *digital* and *analog*. These terms are intended to indicate that the brain handles the same types of information that digital and analog computers do, not that it necessarily does it the same way. Analog information is processed by looking for similarities and differences (i.e., analogies) between various information sets, while digital information is processed by manipulating symbols according to predetermined rules. Analog operations are *qualitative* in nature and deal primarily with patterns and pictures, while digital ones are *quantitative* and deal mostly with language and numbers. Digital information processing is concerned with what statisticians call *discrete* variables, things that can be individually counted, like the number of jellybeans in a jar or elephants in a herd (Moody 2004). Analog information processing is concerned with *continuous* variables, things that can only be measured relative to some sort reference of standard, like length, weight, temperature, and blood pressure. Digital information can be processed by mathematical and logical operations that enable it to be added or subtracted from other digital elements to form new informational patterns. Analog information can only be expressed in relation to a particular reference pattern, so that, although we can say that something is larger than a breadbox or faster than a speeding bullet, we cannot add or subtract this information from something else.[2]

The two ways our brains process information are similar in some ways to the two ways the devices we have invented process it. Analog computers compare information patterns that are fed into them with ones that have been set previously, while digital computers perform logical and mathematical operations on discrete "bits" of input data according to previously specified rules. Analog computers use electrical, mechanical, or hydraulic phenomena to model the problem being addressed, while digital ones manipulate coded symbols, such as a string of ones and zeros. The difference between digital and analog information can readily be seen in the different ways we represent time: either by discrete numbers (e.g., 3:30 p.m.) or by the continuous movement of hour and minute hands against a reference background (e.g., ◐). The difference is also reflected in the way we capture photographic images by either stimulating light-sensitive chemicals on film in an analog camera or activating electronic cells in a digital one. Although digital methods of representing information have replaced analog ones in many of the devices we use, a number of analog instruments are still in use, including blood pressure machines, mercury thermometers, and automobile speedometers.[3]

Digital information-processing systems have three main advantages over analog ones: (a) their discrete "bits" can be combined in a profusion of different ways to represent new entities and relationships, (b) they can be programmed for a number of different applications (while analog ones cannot), and (c) they can represent information portrayed by symbols. Symbolic information processing is, of course, what enables us to communicate with one another through language and use numbers to quantify items and make calculations. Analog information can be converted into digital information, however, by devices that arbitrarily divide it into discrete segments that can be represented numerically, like the way an analog image can be converted into a digital one by splitting it into a number of tiny squares or "pixels" (short for picture elements).

Discrete Objects

Although there are no naturally occurring discontinuities in the stream of perceptual information that our waking brains receive, infants soon learn to create them and identify discrete objects in the world around them. They begin to distinguish specific people and specific objects as they try to make sense out of the swirl of sensory stimuli that impinges on them. Over time, particular patterns of sensory stimulation become recognized as meaningful information and, as this happens, they begin to experience a degree of order in the "blooming, buzzing confusion" that surrounds them (James 1890, p. 462). Cognitive development in young children involves transforming sensations into meaningful perceptions, and then, as their brains continue to develop, understanding the relationships that link these together (Gopnik et al. 1999).

While the patterns our chemical receptors identify are predetermined by the way that particular molecules fit into the proteins that "recognize" them, the meaningful patterns our visual and auditory systems detect are largely learned from the way we experience and understand the world. Infants initially learn to identify discrete objects by noting which elements remain constant and which change when they compare their current sensory input with the input that immediately preceded it. Elements that always seem to cluster together and move in unison are perceived as identifiable entities that can interact with one another, while those that do not form an undifferentiated backdrop against which such interactions take place. Our visual system recognizes discrete objects by detecting distinctive features among the pattern of incoming neural impulses, and then reassembles these to form the perceptions we actually experience. Certain shapes and features are recognized as being similar to previously encountered ones, probably because they trigger a similar pattern of neuronal activity (Roederer 2005, p. 202). The process becomes progressively more discriminating as an increasing number of individual objects are identified. Particular objects and events come to acquire distinct meanings and generate distinct responses, so that apple comes to trigger a different response from banana, and Fido a different one from Fluffy.[4]

The properties of the objects we perceive and the relationships between them are *inferred* by the brain, rather than detected by our sensory receptors, since they are mental constructs that have no material existence of their own. Large and small, rough and smooth, before and after, for instance, are not explicitly represented in the receptor patterns that objects and events activate, but are inferences we make about the relationships we perceive. We locate most of the properties that we attribute to objects on a continuum that allows us to estimate their relative strength or intensity, such as the ranges we perceive between extremely fast and extremely slow and extremely bright and extremely dull. We also link related objects together as associations in our memories, so that experiencing one evokes the memory of the other. Our brains build up mental representations of the external world and how it works out of a mixture of inference and experience, and we then use the models this creates to understand new information.

Every time a new object is identified, a new pattern of connectivity is established between the receptors and effectors that it activated. If these connections are subsequently damaged by injury or disease, individuals become unable to detect meaningful information or remember what they previously learned. Patients with Alzheimer's disease and other dementias gradually lose their cognitive abilities as their neural circuits become disrupted, until they eventually fall prey to the same sort of confusion that infants initially experience. Although visual and auditory stimuli continue to impact their sensory receptors, affected individuals become unable to distinguish meaningful signals from meaningless noise, and thus can no longer make sense of the world. Their perceptions of smell, touch, and taste are less affected, however, possibly because the analogic circuits that process them contain fewer synapses than the digital ones involved in vision and hearing, and are thus less exposed to being disrupted.

Wholes and Parts

The notion that the whole contains more than the sum of its parts is the underlying principle of *holistic* ways of thinking. Even though most people know that a living organism cannot be understood simply by analyzing its chemical components, or a television set by examining its electronic ones, holistic principles have been difficult to operationalize. Gestalt Psychology and Systems Theory, the two main disciplines built on holistic principles, have met with only limited success in real-world situations, especially when compared to the achievements made by reductionistic approaches that take things apart to see how they work (Von Bertalanffy 1968, p. 54). One of the main difficulties has been finding a viable framework for capturing what is so unique about *wholes*; what is it about salt, for instance, that cannot be predicted from understanding sodium and chlorine, or about the functioning of the brain that cannot be predicted from simply understanding its neuronal structure. One explanation is that the *form* an entity assumes, the way its parts are arranged, can have an effect that is independent of the ones the parts themselves have. These are often called *emergent properties* because of our inability to predict them from studying an entity's component parts (Von Bayer 2004, p. 151). Although emergent properties can be physical in nature, like the surface tension of water, many of them are informational and can only be sensed by living beings.[5]

Another approach to understanding holistic properties is to see them as a function of the way an entity relates to the world around it, that is, to the larger system of which it is itself a part. There are, of course, no true wholes and parts in the universe, since every whole is a part of some larger entity, and every part is a whole to the components that make it up. As Hawkins and Blakeslee (2004, p. 126) point out: "All objects of the world are composed of sub-objects that occur consistently together; that is the very definition of an object." Because such a seamless universe offers no handles to grab hold of, we carve it into arbitrary units and examine how their component parts interact. If, however, we want to understand *why* things function the way they do, rather than *how* they do, we need to examine the effects they have on the external entities with which they interact. Questions like why do rattlesnakes rattle, zebras have stripes, or fish swim in schools, can only be answered by examining the impact these features have on the world in which these animals live. Although we can act as if the brain is a closed system (i.e., something complete in itself) in order to learn how it operates, we need to examine the way it interacts with other entities (i.e., as part of an open system) if we want to understand why it does what it does. Thus, while the reductionistic (mechanistic) approach deals with the brain's anatomy and physiology, the holistic one addresses its cognitive, emotional, and behavioral operations. These two approaches have to be reconciled in some fashion, however, if we want to understand how the brain's structures and functions are related (Bohm 2002, p. 11).[6]

Symbols

A symbol is a discrete pattern of energy or matter that uniquely represents some other entity, conveys the same information and meaning, and generates a similar response. There are two major types: *iconic* symbols, which are related is some way to the entities they represent, and *coded* symbols, which are arbitrary tokens that have no logical connection to what they stand for. Iconic symbols include both visual and auditory patterns, like religious emblems and national anthems. The meaningful information that iconic symbols convey is processed analogically and cannot be rearranged to form new representations. Coded symbols, on the other hand, are processed digitally and can be combined to build new informational patterns, much like pieces of *Lego* can be linked together to form novel structures. They are usually part of an organized system in which a unique symbol is designated by a preassigned code to represent each involved entity. Coded symbols are the basic informational units used in language and digital computers, and probably as well as in the mental models we construct to represent the world we know.

Every one of the various spoken and written languages has its own unique system of assigning symbols to encode the sounds and images it employs—as also presumably does the neural language the brain utilizes to process them. All that is required is that there is a separate and distinct pattern of sounds, images, or neuronal connections for every object, property and relationship represented. The arbitrary nature of the symbols different languages create is evident in the differing arrays of the sounds and shapes they use to convey information. The disconnection between coded symbols and what they represent also characterizes other forms of symbolic information, including the bar codes on supermarket products and the strings of ones and zeros that activate digital computers. Coded symbol systems are governed by sets of rules that both senders and receivers have to know in order for them to convey meaningful information.

The special skills that allow us to process complex symbolic information are probably our greatest evolutionary achievement. They are the basis of most of our unique cognitive abilities, including the way we use language, reason logically, imagine the future, and experience a sense of self. Although some hand-reared animals have been trained to detect and respond to a number of distinctive symbols, they have not been able to process the syntactic structures that characterize human language. While Kanzi, the pigmy chimpanzee raised by Sue Savage-Rumbaugh, and Alex, the gray parrot raised by Irene Pepperberg, learned to respond to several hundred words and numbers, these skills appear to be based more on analog processes than on digital ones (Savage-Rumbaugh and Lewin 1994; Pepperberg 1999). Human infants, on the other hand, acquire the ability to process symbolic information during their first few years, so that by the time they are 3-years old they are usually able to combine words into sentences, recall past events, and anticipate future ones.

Unlike the things they stand for, verbal symbols can readily be manipulated and rearranged to convey endless amounts of new information. The digital information that our brains process is not limited to language functions, however, since anything that can be perceived as a discrete object, either real or imaginary, can be represented by a neural symbol. We can, as a result, rearrange our living room furniture in our mind or review career options in our imagination without having to do either in the real world. The representational models the brain constructs are made by linking clusters of neural symbols together and capturing the relationships between them. These models are not miniature replicas of the entities involved, but consist instead of a variety of maps, associations, theories, and representations that depict the causal and structural relationships that link their component elements together—and explain how they function. The models we construct in our mind are not entirely unlike the ones we create in the real world, since both usually contain only enough of the relevant variables to make probabilistic predictions, rather than exact ones (Keller 2002).

Neural Symbols

While no one knows how the brain creates the symbols it uses to portray concrete objects or represent their properties and relationships, it likely has some resemblance to the way it processes language. The cortex apparently uses patterns of neuronal connectivity to encode the various symbols it processes, in much the same way as we use patterns of matter and energy to encode the words we use. The translation of visual and auditory symbols into neural ones involves the conversion of information patterns in one medium into comparable patterns in another, like the way a computer converts the words and numbers we type on its keyboard into electronic strings of ones and zeros. As long as the two configurations are isomorphic they can represent the same information—that is, as long as their components are arranged in a comparable, nonrandom arrangement in which every part of one is mapped onto a corresponding part of the other.

The way that digital computers use combinations of on/off switches to process and store information provides a way of thinking about how the brain might perform similar functions. The binary code that computers use is based on a series of 8-digit arrays of ones and zeros, with each digit being called a *bit* (binary digit) and each 8-bit unit a *byte*. Each of the ones is encoded as "on" and each of the zeros as "off" in the computer's electronic circuits, so that a byte can represent a particular letter or number by a distinctive pattern of its on/off switches. The binary code for the number 184, for example, is 10111000, and for the letter R is 01010010, each of which can be converted into a unique pattern of electronic activity. Computers have circuits that can add, subtract, multiply, and divide these numerical symbols, as well as link them together to represent objects and form words and sentences, all according to previously programmed rules.[7]

Although the system of symbols the brain uses to process digital information is still a mystery, the way neurons interact with each other enables us to consider how it might operate. Since each of the various signals a neuron receives from other neurons merely *helps* to excite or inhibit it, it does not fire a signal of its own unless the sum of its various inputs reaches a critical level. McCullogh and Pitts (1943) were the first to point out that, because neurons respond in this all-or-none manner, they could function as binary on/off switches and represent information in the brain in a similar way to how it is represented in computers. The manner in which neuronal signals can be activated or inhibited by the input from other neurons also allows them to perform the same logical *and*, *or*, and *not* operations that digital computers use to capture different types of relationships (Johnson-Laird 2006, p. 107). The notion that brains might function like computers also meant that computers might function like brains—and could be used to simulate their functions.

The brain processes analog information more rapidly than digital information, which is the reason we can tell the time faster on an analog watch than a digital one, and why recognizing a face is much easier than describing it—for, as they say, a picture is worth a thousand words (Paivio 2006, p. 35). Analog devices are not as flexible as digital ones, however, since they are limited to performing specific tasks, like letting us know how fast we are traveling or how much gas is left in the tank. Their advantage is that any number of them can operate simultaneously (i.e., in parallel), which is apparently how various parts of the brain function. The brain processes information much slower than a computer, however, since signals can travel a million times faster on a microchip than over a neuronal pathway (Carruthers 2006, p. 24). It makes up for this, however, by performing a host of different analog comparisons simultaneously. As Morgan (2003, p. 44) notes: "The slowness of the brain would be a devastating disadvantage, if it were not for the fact that millions of nerve cells can be active at the same time, performing different computations." Much of the uniqueness of the human brain lies in how it is able to integrate its digital and analog information-processing systems in ways that current computers are unable to match.

Modality Differences

Vision and hearing are unique among our sensory modalities because they can detect and process both digital and analog types of information. Our senses of taste, smell, touch, and kinesthesia are ordinarily limited to processing analogic patterns (although Braille is one exception). While our direct, moment-to-moment experience of the world is processed by means of analog comparisons, most of our distinctive cognitive functions rely on digital-type operations. These not only allow us to process language, but also to set aside our current sensory input in order to analyze what is going on, make decisions, and plan for the future. As Donald (2001, p. 157) notes: "Humans bridge two worlds. We are hybrids, half analogizers, with direct experience of the world, and half symbolizers, embedded in a cultural web. During our evolution, we somehow supplemented the analogue capacities built into our brains over hundreds of millions of years with a symbolic loop through culture."

The representational models formed in the brain are made up of configurations of neural symbols that do not resemble the things they represent in any recognizable way, although they contain the same information. We are able to construct or recall scenarios in our mind by linking encoded visual and auditory representations into patterns that portray the imagined events. We are not able, however, to recall and re-experience analogically processed sensations in the same way, even though we can recognize that a particular taste or smell (or touch or pain) is similar to one we experienced previously (Denton 2006, p. 178). We also have a difficult time finding words to describe the feelings our analog-based senses generate, presumably because language involves digital processing. Taste, smell, and tactile sensations are also absent from our worlds of fantasy and imagination, since these simulated realms are created from the brain's store of digital symbols and representations. Even the hallucinations experienced by individuals with schizophrenia are typically limited to visual and auditory perceptions, since tactile, taste, and olfactory hallucinations are usually caused by structural brain lesions.

Notes

[1] Shettleworth (1998) discusses research findings about the cognitive abilities of humans and other species from both a psychological and an evolutionary perspective.
[2] The Josiah Macy, Jr. Foundation organized a series of ten conferences between 1946 and 1953 to which the leaders of the emerging fields of neuroscience, cybernetics, information theory, and cognitive psychology were invited. A summary of the proceedings is available at http://www.

asc-cybernetics.org/foundatuons/history/MacySummary. Ralph Gerard, the co-founder of the Society for Neuroscience, commented at the initial meeting that the brain's operations are much more analog than digital, and started the seventh conference with a presentation on analog versus digital interpretations of the mind. Although the idea that the brain might process information analogically stimulated animated debate at the conferences, it seemed to fade from view after they were over.

[3] Miller et al. (1960) note: "In the language of computing machines, an analog device is one in which the magnitudes involved in the computations are represented by physical properties proportional to these magnitudes, i.e. by a voltage, a duration, an angle of rotation, etc. A digital computer, on the other hand, represents the magnitudes by which it works by symbols corresponding to discretely different states of the machine. There is no simple resemblance between the input to a digital computer and the processes that represent that input inside the machine." Analog computers were used by a number of scientific projects during the first half of the twentieth century to match various inputs against mechanical or electrical models. The first programmable (i.e., digital) computers were introduced in the early 1940s, but these were bulky, impractical machines that depended on hundreds of unreliable vacuum tubes. They were replaced in the 1960s by transistor-based devices, which subsequently became ever smaller, faster, and cheaper as the technology on which they were based evolved (Mindell 2002).

[4] Certain visual and auditory cues are able to evoke innate responses, like the smiling face or cry of a baby, but this is not the primary way we acquire information about the world. The neural substrates that enable us to acquire language and learn about the world are also part of our genetic heritage, even though the languages themselves are learned (Chomsky 1972).

[5] Capra (1996, p. 17) reviews the interplay between holistic and reductionistic concepts in Western science from Aristotle and Descartes to current System Theories. Systems Biology is a relatively new field that aims at understanding interactions between the components of biological systems, although its focus has been on understanding physical interactions, not informational ones (Kitano 2001). Although analysis is a necessary aspect of science, it does not mean that proceeding to ever-finer levels of discrimination is necessarily helpful. As Mayr (2004, p. 69) points out, understanding the component parts of a hammer or a cell up to a point helps understand their function, but proceeding further to the level of atoms and electrons does not shed any additional light on them.

[6] According to the second law of Thermodynamics, closed systems increase in entropy and become more disorganized over time. Open systems avoid this and maintain their integrity by allowing the transfer of energy and matter across their boundaries. While living entities function as open systems, inanimate ones do not, which is why they gradually decay and wear out.

[7] Claude Shannon's master's thesis at MIT in 1937, *A Symbolic Analysis of Relay and Switching Circuits*, created the practical design of digital circuits by using electronic relays and switches to represent binary arithmetic (ones and zeros) and Boolean algebra (and, or, not). Shannon (1916–2001) is also credited with developing *Information Theory* with his 1948 paper on *A Mathematical Theory of Communication* (see Chap 17), as well as with introducing the sampling theorem that allows analog information to be converted into digital electronic signals.

References

Bohm D (2002) Wholeness and the Implicate Order. Routledge, New York

Capra F (1996) The Web of Life: A New Scientific Understanding of Living Systems. Doubleday, New York

Carruthers P (2006) The Architecture of the Mind: Massive Modularity and the Flexibility of Thought. Oxford University Press, Oxford

Chomsky N (1972) Language and Mind. Harcourt Brace Jovanovich, New York

Denton D (2006) The Primordial Emotions: The Dawning of Consciousness. Oxford University Press, New York

Donald M (2001) A Mind So Rare: The Evolution of Human Consciousness. WW Norton, New York

Gopnik A, Melzoff AN, Kuhl PK (1999) The Scientist in the Crib: Minds, Brains, And How Children Learn. William Morrow, New York

Hawkins J, Blakeslee S (2004) On Intelligence. Henry Holt, New York

James W (1890) Principles of Psychology. Henry Holt, New York

Johnson-Laird PN (2006) How We Reason. Oxford University Press, New York

Keller EF (2002) Making Sense of Life: Explaining Biological Development with Models, Metaphors, and Machines. Harvard University Press, Cambridge

Kitano H (2001) Foundations of Systems Biology. MIT Press, Cambridge

McCullogh WS, Pitts W (1943) A logical calculus of the ideas immanent in nervous activity. Bull Math Biophysics 5: 115–133

Mayr E (2004) What Makes Biology Unique? Considerations on the Autonomy of a Scientific Discipline. Cambridge University Press, Cambridge

Miller GA, Galanter E, Pribram KH (1960) Plans and the Structure of Behavior. Holt, Rinehart & Winston, New York

Mindell DA (2002) Between Human and Machine: Feedback, Control, and Computing before Cybernetics. Johns Hopkins University Press, Baltimore

Moody G (2004) The Digital Code of Life: How Bioinformatics is Revolutionizing Science, Medicine, and Business. John Wiley & Sons, New York

Morgan M (2003) The Space Between Our Ears: How the Brain Represents Space. Weidenfeld & Nicholson, London

Paivio A (2006) Mind and Its Evolution: A Dual Coding Theoretical Approach. Lawrence Erlbaum, Malwah

Pepperberg IM (1999) The Alex Studies: Cognitive and Communicative Abilities of Grey Parrots. Harvard University Press, Cambridge

Roederer JG (2005) Information and Its Role in Nature. Springer, New York

Savage-Rumbaugh S, Lewin R (1994) Kanzi: The Ape at the Brink of the Human Mind. John Wiley & Sons, New York

Shettleworth SJ (1998) Cognition, Evolution, and Behavior. Oxford University Press, New York

Von Bayer HC (2004) Information: The New Language of Science. Harvard University Press, Cambridge

Von Bertalanffy L (1968) General System Theory. Foundations Development Applications. George Braziller, New York

Chapter 8
Storage and Retrieval

Abstract Information is represented in the brain as patterns of synaptic connections within and between arrays of linked neurons, and this determines how we interpret, remember, and respond to what we encounter. Memory and learning are indistinguishable at the neuronal level, since both involve similar changes in the synaptic connections that link nerve cells together as a result of experience. Recognition is the sense of familiarity elicited when a current experience activates the memory of a previous one; recall is the re-experiencing of previously perceived objects or events. Recalled memories are reconstructed, however, not just re-elicited. Although meaningless sensory stimuli can be experienced consciously and temporarily retained in working memory, only meaningful information gets stored in long-term memory.

Although most neuroscientists believe that meaningful information is stored in the brain as specific patterns of neural connectivity, no one knows exactly how it is done. The idea is that information is represented as patterns of synaptic connections within and between arrays of linked neurons—and that these stored configurations help determine how we interpret, remember, and respond to what we subsequently encounter. Meaningful information can become encoded in the brain through three main routes, each of which leaves its imprint in a different way. First, information in our genes shapes the brain's overall structure and its strategy for further development, as well as the initial patterns of synaptic connections responsible for innate predispositions, reflex responses, and homeostatic regulation. Next, the perception of external objects and events generates specific patterns of neural connectivity that enable familiar entities to subsequently be recognized and recalled. Finally, language and other forms of symbolic information are stored as digital representations that can be organized and manipulated to portray inferred relationships, abstract concepts, and imagined scenarios (Bourtchouladze 2002).

The human brain is able to store immense amounts of information because of the endless patterns with which its 100 billion neurons can be connected to each other.

A. Reading, *Meaningful Information: The Bridge Between Biology, Brain, and Behavior*,   63
SpringerBriefs in Biology 1, DOI 10.1007/978-1-4614-0158-2_8,
© Springer Science+Business Media, LLC 2011

Our brains have little trouble, for instance, in accommodating the complex knowledge and cultures that have accompanied modern science and technology, even though their capacity for storage evolved during much simpler times. Individuals are able to acquire huge amounts of information throughout their lifetimes, both formally through the education system and informally from their own experiences and contacts. The number of familiar objects, images, sounds, facts, and sensations we can recognize seems limitless, even though we are not able to recall many of them without prompting. Our species is also unique in the way we have developed external storage devices that supplement our neuronal abilities, like books, pictures, libraries, and the Internet.

The human brain's vast storage capacity is clearly evident in the feats of *savants,* like the one who inspired the 1988 movie *Rain Man.* This remarkable individual could remember more than 7,600 books by heart and recite the highways that go to each American city, town, or county, as well as their area and zip codes, television stations, and telephone networks. He could instantly tell a person who gave him their birth date what day of the week it fell on and what news items were on the front page of the major newspapers that day. He could also identify most classical compositions and recall the date the music was published or first performed, the composer's birthplace, and the dates they were born and died (Treffert and Christensen 2005). Some savants have true photographic memories that enable them to picture an entire page at once, like a camera does, so that they can still read from it when they turn their gaze elsewhere. Despite their amazing talents, these individuals usually do not understand the meaning of what they remember, but just repeat back what they have recorded. Most of them have problems functioning independently, partly because their focus on detail interferes with their ability to form abstract concepts and generalize from what they learn. It is not yet clear, however, how their brains differ from those of more normally functioning individuals (Nettelbeck 1999; Treffert 2006).

Memory and Learning

Memory and learning are virtually indistinguishable at the neuronal level, since both involve similar changes in the synaptic connections that link nerve cells together as the result of experience. Learning allows us to respond in a new way when exposed to a situation we have previously encountered, while memory makes it possible for us to recognize or recall previously perceived objects and events. Motor skills and other forms of knowledge that are learned outside of conscious awareness, like those involved in riding a bicycle or knitting a sweater, are stored in what is called *implicit memory,* since they are not directly accessible. It seems that the only type of stored information we can access consciously is information that was perceived consciously, although we may need prompting to help retrieve it. Seeing an old photograph or meeting an old friend, for instance, can release memories that we had completely forgotten. We are generally unable to remember our dreams, retrieve what we learned through imitation and identification, or recall events that happened before we acquired symbolic language. Stored information that can be accessed consciously is referred to as *explicit memory*. This is comprised of both *episodic memory,* which stores our individual life experiences, and *semantic memory*, which stores the symbolic information we acquire through spoken and written language (Tulving and Craik 2000).[1]

Episodic memory forms a record of the sensory events we have personally experienced—the meaningful sights, sounds, smells, tastes, and feelings that have filled our daily lives. It allows us to recognize whether new perceptions are similar to previous ones and distinguish familiar objects from unfamiliar ones. This is accomplished through an analogic process that matches newly created patterns of neuronal activity against previously generated ones, with the sense of familiarity or unfamiliarity we feel being the output of these comparisons. No one knows just how incoming information is able to find the relevant memory templates from among the billions of neural possibilities, or how the feelings of familiarity or unfamiliarity are generated.

Semantic memory involves the storage of language-related symbols. It consists essentially of facts, rules, and relationships that have either been acquired from our own experience or communicated to us by others. Verbal information not only enables us to have access to other people's experiences, but also to the knowledge and understanding they have derived from them. The verbal symbols that make up language are encoded in the brain as neural symbols that are then linked together to represent various objects and events, as well as their properties and relationships. Whereas episodic memory stores instances of tangible entities, like the image of Aunt Louise or the Statue of Liberty, semantic memory stores facts *(4 × 6 = 24),* relationships *(pride comes before a fall)*, and abstract concepts *(justice, friendship)*. The symbols encoded in semantic memory are processed digitally, while the pictures and patterns in episodic memory are processed analogically.

Implicit memory is comprised of the types of information we acquire automatically, without necessarily being aware of it. It includes the neural linkages responsible for acquired motor skills, the associations formed through conditioned learning, and knowledge obtained through subliminal perception. We also learn many of our values, attitudes, and prejudices implicitly by imitating and identifying with role models without being aware we are doing it, which is probably why they are so resistant to change. While explicit memory appears to be limited to information stored in the cerebral cortex, implicit memory includes patterns formed in the cerebellum and sub-cortical regions of the brain.

Recognition and Recall

Recognition and recall are the two ways we access information stored in memory (Brown 1976). *Recognition* is the sense of familiarity elicited when a current experience activates the stored record of a previous one, while *recall* is the conscious re-experiencing of previously perceived objects and events, either cued by a current stimulus or elicited without external prompting. Recognition is based on a matching process that compares current sensory input with information patterns that have been stored in the brain, while recall is based on the reconstruction of previously experienced information. Recognition is the more basic of the two, since it encompasses all of the different sensory modalities and is present throughout the animal kingdom. Recall is a predominantly human experience, especially when not cued by external events (Tulving 2005).[2]

As discussed in the previous chapter, we are normally unable to recall past sensations of taste, smell, touch, or pain. Although we can recall we had them, as well as the circumstances in which we did, we cannot re-experience the sensations themselves in the absence of a current stimulus. Vision and hearing are the only modalities we can freely conjure up in our imagination, either to recall scenarios from the past, imagine ones about the future, or concoct daydreams and fantasies. However, the images and sounds we create are not as bright or clear as the ones we experience directly, nor do they contain the same amount of detail. Recalled memories are prone to a variety of errors, even though individuals are usually adamant about their accuracy. Most psychologists now believe that false memories occur because recalled experiences are reconstructed from stored perceptual elements, rather than re-elicited from a fully intact record of past events. As Gilbert (2006, p. 197) points out: "Memory is not a dutiful scribe that keeps a complete transcript of our experiences, but a sophisticated editor that clips and saves key elements of an experience and then uses these to rewrite the story each time we ask to reread it."[3]

Recalled memories can be distorted by input from a number of extraneous sources, including preexisting beliefs, external suggestions, and overlap with other memory traces (Loftus 2000). Even the vivid "flashbulb" memories people have of what they were doing when they first heard the news of a catastrophe are not as indelibly inscribed as originally thought. Neisser and Harsch (1992), for instance, report that students who were asked to record their memories of the Challenger disaster a day after it happened, and again 2–3 years afterwards, had a surprising number of errors in their later recollections, even though they were confident of their accuracy. As Dowling (2004, p. 2) observes: "Contrary to the long-held belief that an eyewitness can faithfully record and remember an event, we now realize that what we remember or even perceive of an event depends on many factors—previous experiences, biases, attention, imagination, and so forth." Errors seldom occur in recognition memory, although they do turn up occasionally as either *déjà vu* or *jamais vu* experiences (i.e., feeling that an unfamiliar scene is familiar or that a familiar one is unfamiliar).

Our capacity for recognition is much greater than our capacity for recall. We can recognize a seemingly endless profusion of the things we have previously seen or heard or smelled or tasted, although we cannot necessarily recall them. A number of studies have documented the extent to which we are able to recognize objects more readily than recall them. The subjects in one were able to accurately recognize 90% of 2,500 color slides they were shown for 10 s each 10 days previously (Broadwell 1994), while those in another correctly identified 66% of 10,000 pictures they were shown briefly 2 days earlier (Standing 1973). In neither study, however, were the participants able to independently recall any of the items. The differences between recognition and recall are most likely due to the different ways they are processed, since recognition involves detecting only a few key features of an event, while recall entails reconstructing the complete episode.

Semantic memories also differ from episodic ones in that they contain information about inferred properties, like size, weight, and texture, as well as about spatial, temporal, and causal relationships. The properties and relationships that become represented in our nervous systems are not sensed directly, however, for they have no material existence of their own. They are simply abstractions we construct to explain how the objects and events we detect interact with each other and with us. Memory in most other animals is not a conscious activity, but just their way of retaining the effects that meaningful information has had on them. Even immune cells and unicellular organisms learn and remember, since both are able to change their behavior as a result of experience.

Retention

The objects and events to which we are currently paying attention are temporarily retained in working memory, and then either discarded or transferred to long-term storage. Working memory involves retaining the pattern of neuronal activity generated by the incoming sensory input for a few seconds, so that the brain can detect if there are any unexpected discrepancies between it and the pattern that immediately follows it. Keeping sensory input in working memory also gives the brain time to determine whether any meaningful information is embedded within it. These meaningful patterns are the only part of working memory that ends up being stored in long-term memory, where they provide the substrate for the individual's subsequent recognition and recall activities. Although long-term memories involve synthesizing new proteins and forming new synaptic connections, they can apparently be maintained without the active energy supply that working memory requires. (Roederer 2003, p. 19). This is why neural traces of currently inactive memories do not show up on functional brain scans.[4]

No discernable material is added to the brain during learning and memory, just as none is added to the hard disc of a computer as it stores new data. Memories are weightless, even great ones, since they only involve changes of the organizational patterns in the brain, not changes in its size or substance (Niele 2005). Memory does not appear to be stored in a separate location in the brain; it is just a reflection of the way the neural activity generated by the perception of meaningful events is retained. This is accomplished by converting the temporary patterns of working memory into more lasting connections between the involved neurons. Information in the brain is not stored *in* memory, like the way it is in a computer, but is stored *as* memory, since nothing is actually *stored* anywhere during the process. Memory is simply a record of the effect that the perception of meaningful information has had on the brain's patterns of neural connectivity, as both the perception and the recall of an object activate the same parts of the brain (Galaburda et al. 2002).

What Gets Stored

While meaningless sensory stimuli can be experienced consciously and retained temporarily in working memory, they do not get stored in long-term memory. Although we ordinarily witness the events that parade before us in their entirety, we only remember those features that change our behavior, physiology, or neural structure (i.e., that are meaningful to us). How meaningful experiences are depends largely on whether they elicit an emotional response, while how meaningful facts are depends on how closely they fit with what the particular recipient already knows, since this enables them to be interpreted within a preexisting framework. Isolated facts that have nothing to anchor them to the recipients' interests or knowledge are difficult to retain, which is why they tend to "go in one ear and out the other" without having any effect on the individual. Although we have some awareness of the peripheral objects in our field of vision as we drive along the highway or watch a football game, we only retain the part of our perceptions that had a meaningful effect on us. This is why we only remember the *gist* of what was said when we read a book or listen to a conversation, not the actual words, for it is the only aspect that has a meaningful effect on us.

The minimum number of features needed to elicit a response to a perceived object is usually all that is retained for later reference, for this is all it takes to identify meaningful information. We only have to recognize a few key features, for instance, to identify a particular face or voice or tell whether an animal is a cat or a dog. Whether someone or something is recognized as an individual or as a member of a class or category, however, depends on whether the distinction makes a meaningful difference to the observer. If one cow is as good as another to an individual, the key features of a generic cow are all that need to be detected, but if the observer is a farmer who differentiates between *Betsy* and *Lulubelle*, then some of their individual features also need to be stored.[5]

Although episodic memory constitutes a record of the objects and events that have been meaningful to the individuals involved, it apparently does not record all of their features every time they encounter them. We do not create a new record every time we see a family member or a familiar pet, but modify our core template of them to fit the particular circumstances. Because we do not lay down a continuous record of what we see, the way a video camera does, the images we later recall are not exact replicas of what was previously experienced. Recalled scenes are apparently constructed the same way imagined ones are—by combining their key features with other stored information that fills in the missing details. As Damasio (1994) notes: "Whenever we recall a given object or face or scene, we do not get an exact reproduction, but rather an interpretation, a newly reconstructed version of the original."

Paying Attention

We can only give our full attention to one part of our sensory environment at a time. We do this with vision, for instance, by focusing our gaze, so that the light from a particular entity falls on the fovea, the small area near the center of each retina that processes high-resolution details. The features detected by the fovea's receptors are the only ones that get analyzed and compared to stored representations, since the receptors outside this area are less able to discriminate between meaningful and nonmeaningful patterns. Thus, although sensory inputs from other parts of the environment are still registered, the section of our visual scene that we are paying attention to is the only one that is actively processed and retained in memory. As Triesch et al. (2003) note: "in everyday tasks only a very limited amount of visual information is "computed" at each fixation—just enough to solve the current sensorimotor micro task."[6]

Several studies have shown that we are relatively blind to certain aspects of our visual environment, despite our sense that we see everything in our line of sight. In one dramatic demonstration, half of the observers who were paying close attention to a videotaped basketball game failed to notice a person wearing a gorilla suit walk into the middle of the game, stop to face the camera and thump its chest, and then walk off after being in clear view for a total of 9 s (Simons and Chabris 1999). In other studies, subjects fail to detect large changes in a scene they are watching if their vision is temporarily disrupted while the change takes place. Many people, for instance, fail to notice when a person they are talking to is surreptitiously replaced by a different individual during a brief period when their view is interrupted (Simons and Levin 1998). People who are engaged in engrossing tasks can fail to notice obvious events taking place outside their immediate sphere of attention. This occurs even when the competing information is on a different sensory channel, like someone being distracted from driving by talking on a cell phone or texting.[7]

Notes

[1] Dudai (1989) and Squire (1987) review the neurobiology of learning and memory.

[2] It is not clear whether other primates are able to recall previous experiences, since we have no access to their mental states. However, they do not seem to re-experience past events or create imaginary ones, either in captivity or in the wild. Non-primates are unlikely to be able to recall the past or imagine the future because they are usually locked into the present. Clayton et al. (2003) maintain, however, that the way certain birds remember where they have previously hidden food caches shows that that they are able to recall past events.

[3] Hobson (1999) believes that some people can recall previously experienced odors, mostly in association with images of the entities that generate them.

[4] The hippocampus plays a critical role in the transformation of information from short-term to long-term memory. People with bilaterally damaged hippocampal regions are unable to convert working memories into long-term ones, although they can recognize and recall information that was stored prior to the damage. HM was a well-studied patient who had both of his hippocampal regions damaged as a result of surgery for severe epilepsy. Although his intellect and personality

remained unchanged after the operation, he was no longer able to lay down new memories. He read the same magazines over and over without realizing he had read them before, could not recognize familiar people, and had no idea of what he had just been doing. Because he was unable to remember anything new for more than a few seconds, everything always seemed new to him. He could, however, recall events from his earlier life, showing that long-term memories are not *stored* in the hippocampal region (Scoville and Milner 1957).

[5]Facial features are the primary way that humans and other primates identify members of their species. Freiwald and Tsao (2010) show that facial perception in the macaque involves a sequence of specialized neuronal areas in which the successive ones increasingly respond to individual features.

[6]We make up for the lack of detail in the rest of our visual field by continuously scanning it with extremely rapid eye movements, called saccades, which imperceptibly focus the fovea for an instant on other areas of the environment. If something unexpected is detected in the periphery, we automatically move our eyes to focus our attention on it, so we can analyze and process it.

[7]These effects are called *inattentional blindness* and *change blindness* (Mack and Rock 1998). Magicians exploit them by getting audiences to focus on some distractive element, away from the ones being manipulated. Case studies of these functional types of blindness involve reported objects or obstructions that should have easily been noticed, but were not, including drivers running over bicyclists, train engineers plowing into cars, submarines surfacing under ships, and airline pilots landing on other planes. The individuals involved in these accidents reported that they had no idea the object was there and could not explain their failure to see it. Although these examples involve visual attention, focusing exclusively on any of the other sensory modalities or on internal thoughts can produce the same effects. The videotape used in the gorilla experiment can be seen on the University of Illinois Visual Lab website http://viscog.beckman.uiuc.edu/djs_lab/demos.html.

References

Bourtchouladze R (2002) Memories Are Made of This: How Memory Works in Humans and Animals. Columbia University Press, New York

Broadwell RD (1994) Neuroscience, Memory and Language. Library of Congress, Washington

Brown J (1976) Recall and Recognition. John Wiley & Sons, London

Clayton NS, Bussey TJ, Dickinson A (2003) Can animals recall the past and plan for the future? Nature Reviews Neuroscience 4: 685–691

Damasio AR (1994) Descartes' Error: Emotion, Reason, and the Human Brain. GP Putnam's Sons, New York

Dowling J (2004) The Great Brain Debate: Nature or Nurture? Joseph Henry Press, Washington

Dudai Y (1989) The Neurobiology of Memory: Concepts, Findings, Trends. Oxford University Press, Oxford

Freiwald WA, Tsao DY (2010) Functional compartmentalization and viewpoint generalization within the Macaque face-processing system. Science 330: 845–851

Galaburda AM, Kosslyn SM, Christen Y (2002) The Languages of the Brain. Harvard University Press, Cambridge

Gilbert D (2006) Stumbling on Happiness. Alfred A. Knopf, New York

Hobson JA (1999) Consciousness. Scientific American Library, New York

Loftus EF (2000) Remembering what never happened. In E Tulving (Ed), Memory, Consciousness and the Brain. Psychology Press, Philadelphia

Mack A, Rock I (1998) Inattentional Blindness. MIT Press, Cambridge

Neisser U, Harsch N (1992) Phantom flashbulbs: False recollection of hearing the news about Challenger. In E Winograd, U Neisser (Eds), New Affect and Accuracy in Recall: Studies of "Flashbulb" Memories. Cambridge University Press, New York

Nettelbeck T (1999) Savant syndrome: Rhyme without reason. In M Anderson (Ed), The Development of Intelligence. Psychology Press, Hove

Niele F (2005) Energy: Engine of Evolution. Elsevier, Amsterdam

Roederer JG (2003) On the concept of information and its role in nature. Entropy 5: 3–33

Scoville WB, Milner B (1957) Loss of recent memory after bilateral hippocampal lesions. J Neurol Neurosurg Psychiatry 20: 11–21

Simons DJ, Chabris CF (1999) Gorillas in our midst: sustained inattentional blindness for dynamic events. Perception 28: 1059–1074

Simons DJ, Levin DT (1998) Failure to detect changes to people during a real-world interaction. Psychonomic Bull & Rev 5(4): 644–649

Squire L (1987) Memory and the Brain. Oxford University Press, New York

Standing L (1973) Remembering 10,000 pictures. Quarterly J Experimental Psychology 25: 207–222

Treffert DA (2006) Extraordinary People: Understanding Savant Syndrome. Universe, Lincoln

Treffert DA, Christensen DD (2005) Inside the mind of a savant. Scientific American 293(6): 108–113

Triesch J, Ballard DH, Hayhoe MM, Sullivan T (2003) What you see is what you need. J Vision 3(1): 86–94

Tulving E (2005) Episodic memory and autonoesis: Uniquely human? In HS Terrace, J Metcalfe (Eds), The Missing Link in Cognition: Origins of Self-Reflective Consciousness. Oxford University Press, New York

Tulving E, Craik FIM (2000) The Oxford Handbook of Memory. Oxford University Press, New York

Chapter 9
Knowledge and Understanding

Abstract One of the most remarkable things the brain does is transform the raw data of experience into the informational structures that provide a sense of meaning and understanding to our lives. Knowledge is comprised of organized sets of meaningful information that are encoded in the central nervous system, some based on personal experience and some on what others have passed on to us. Understanding is a special form of knowledge that involves appreciating how interacting entities affect each other. Rule-based reasoning involves the sequential processing of discrete bits of information according to strict algorithmic rules, while pattern-based reasoning involves searching for correspondences and contrasts between perceived patterns and reference ones. We organize and rearrange information to build models that reflect our understanding of how individual facts and inferences are linked and interact. We have no way of being certain that the knowledge we accrue is correct, however, since our information-processing systems are not good truth-detectors.

Of all the amazing things the human brain does, one of the most remarkable is the way it transforms the raw data of experience into complex informational structures that provide a sense of meaning and understanding to our lives. This transformation represents an example of *self-organization,* a biological process in which entities change from a less-organized state to a more-organized one in apparent defiance of the laws of physics. The patterns of increased neuronal connectivity that are formed in this process encode the rules, concepts, beliefs, representations, and models that govern our daily lives. The human brain serves both as a library that stores and catalogs the information it receives, as well as a workshop that creates new forms of information from elements it has previously stored. It weaves the neuronal symbols that represent the objects and events we have perceived into new patterns of connectivity that portray their properties and relationships and help us make sense out of what we encounter.

A. Reading, *Meaningful Information: The Bridge Between Biology, Brain, and Behavior,*   73
SpringerBriefs in Biology 1, DOI 10.1007/978-1-4614-0158-2_9,
© Springer Science+Business Media, LLC 2011

Knowledge

Knowledge is comprised of organized sets of meaningful information that are encoded in the central nervous system, some based on our personal experience and some on what others have passed on to us. Knowledge has a number of different forms, including knowing facts, knowing how to do something, knowing how something works, knowing its significance, and knowing how it is related to other things, as well as knowing what it is made of, what it looks or tastes like, where it is located, and how much it costs. Although each of these involves a different set of neural connections, they all fall into one of the two main categories: *explicit knowledge,* which is consciously accessible and derived from consciously perceived information, and *implicit knowledge*, which is derived from genetic programs and implicit memories and not consciously accessible. Explicit knowledge is primarily verbal, like knowing that a cello is a musical instrument or who is buried in Grant's Tomb, while implicit knowledge is usually procedural or attitudinal, like knowing how to use a typewriter or having unconscious biases.

Knowledge can be gained either as a result of the way the brain automatically organizes and arranges the information it stores or through conscious reasoning and analysis. Our brains try to make sense out of the myriad of sensory stimuli that bombard us by searching for patterns of regularity and relationship among them. Things that have similar properties or functions are linked together, as also are ones that occur simultaneously or regularly follow each other. Most of the knowledge structures we form are derived from symbolic representations stored in semantic memory, since these can be combined and rearranged to form new neuronal configurations, while the experiential patterns in episodic memory cannot. Language is the primary vehicle through which we consciously acquire knowledge from others and from the education system, and language-based symbols are the basic ingredients of conscious thought and reflection.

Reasoning

The two main ways we have of consciously creating new informational patterns from existing ones are *rule-based reasoning* and *pattern-based reasoning*. Rule-based reasoning involves the sequential processing of discrete units of information according to strict algorithmic rules, like the way we process language and mathematics, while pattern-based reasoning involves searching for correspondences and contrasts between perceived patterns and reference ones, like the way we process most forms of nonverbal information (Hong 2005, p. 182). We either draw logical conclusions using rules and language-based symbols (i.e., digital processing), or make plausible inferences by comparing spatial or temporal patterns (i.e., analog processing). According to Bruner (1986), we think either in a logico-scientific mode that involves symbols and emotion-free rationality, or in a narrative one that conjures up images and stories, together with the emotions they elicit. Scientists and engineers typically see the world through rule-based lenses, while artists and poets mostly do so through pattern-based ones.

Deduction and induction are the classical ways of deriving rule-based knowledge from previous information, although they are not the only ones (Johnson-Laird 2006). Deduction involves reasoning from the general to the particular through formal logic, as in the classic syllogism: "All men are mortal; Socrates is a man; therefore Socrates is mortal." Induction, on the other hand, entails reasoning from the particular to the general, as in: "All the swans I have known are white; therefore all swans are white." While deductive conclusions are valid if their premises are correct, inductive ones are not necessarily true, even when their premises are, since some swans are black (Taleb 2007). Although philosophers generally look askance at inductive reasoning because of its inherent fallibility, it is how our brains tend to function. Induction is essentially a consequence of the way we organize information into categories, and then act as if the members are identical in all respects (Popper 1959).

Pattern-based reasoning uses the familiar and comprehensible as a way of understanding the unfamiliar and incomprehensible. The process utilizes a model of how one entity functions as the basis for understanding how another does, like envisioning that electricity flows in a wire the way water flows in a pipe, or that electrons orbit around the nucleus of an atom the way planets orbit around the sun. This is the only way we have of initially conceptualizing how an unknown entity operates, which is why scientists used to think that the brain functioned like a telephone exchange. Pattern-based analogies do not have to be perfect to be useful, however, since a poor fit may be better than no fit at all in trying to make sense of the world. They can serve as a first approximation whose discrepancies with the reference system can then be analyzed and used to develop more effective models. Such trial and error approximations can eventually lead to the discovery of new knowledge, since each successive iteration gets the individual one step closer to understanding how the new entity actually functions.[1]

Understanding

Understanding is a special form of knowledge that involves appreciating how interacting entities affect each other. It entails knowing what causes something, what results from it, how to influence it, and how it relates to other phenomena, in other words, having a working model of how it functions (Johnson-Laird 1996). The mental models we build help us understand the world and shape our on-going perceptions by serving as frames of reference for interpreting the meaning of new information. The way we respond to a situation is primarily determined by how we understand it, rather than by its inherent properties, since we interpret the meaning of what we perceive in the light of knowledge we have previously accumulated. This is why people with different backgrounds and different experiences can have such different perceptions of events, and why our supposed objectivity is not always objective.

We develop our understanding of the world both by assimilating new information that is compatible with what we already know, and by modifying what we already know to accommodate information that is not (Piaget and Inhelder 1969). Altering long-held beliefs to fit new information is easier said than done, however, especially when they are entwined with other beliefs that depend on them. We search for clarity and certainty in the universe, even if we have to impose it ourselves by denying or discounting information that conflicts with what we already know. Erroneous beliefs that seem credible to their adherents are almost impossible to dislodge if they serve some type of emotional purpose, like conspiracy theories and End Time prophecies apparently do. The more such distortions become incorporated into people's models, the less accurate these become, and the less able to guide them through the everyday challenges they encounter. The test of how well we understand something is simply how accurately it enables us to predict or control events that either cause it or are caused by it.

Mental Models

We do not just record the information we perceive, but organize and rearrange it to build models that reflect our understanding of how individual facts and inferences are linked and interact. The mental models we develop are critically important knowledge structures, for they play a major role in guiding our daily lives. Whenever we encounter an unfamiliar situation, our brain searches for an existing model that matches it in some way, and then uses this to help us navigate through the new circumstance. The temporal, spatial, and causal models we construct are not, however, miniature replicas of the entities involved, but patterns of neural connectivity that encode the way various aspects of the world are arranged and interrelated. As Craik (1943) observes, a model is "any physical or chemical system which has a similar relation-structure to that of the process it imitates," so that "if an organism carries a 'small-scale model' of external reality and of its own possible actions within its head, it is able to try out various alternatives, conclude which is the best of them, react to future situations before they arise, and utilize the knowledge of past events in dealing with the present and the future."[2]

The working models we develop are like theories, since they are explanatory constructs that portray the relationships that link various aspects of the natural world together. They are also like theories in that they are not necessarily invalidated if events do not turn out entirely as expected, or validated if they do—although they should become suspect if a number of the expectations they generate are wrong (Popper 1959). While real-world theories can gain consensus by being tested in the public arena, our mental models are ours alone to live with and develop, which leaves them open to bias and distortion. Rather than constructing a single, all-encompassing model of the universe, however, we develop a series of interconnected sub-models, each of which represents a different sphere of knowledge. Some of these are shared with other members of our family, social groups, and culture, while others are related more to our individual experiences. It is because of their particular interests, for instance, that pilots know how to fly planes, architects how to design buildings, and ranchers how to raise cattle. Different groups and different cultures build different explanatory models, based primarily on the truths that are most salient to them. People who share the same models are best able to communicate and understand each other, while individuals who develop their own unique belief systems have trouble being understood.[3]

The mental models we develop constitute a private, internal realm where we can conjure up thoughts and images of our own choice, imagine what might be and might have been, set goals for ourselves, and plan how best to accomplish them.[4] Most of us simply assume the world inside our head is the same as the one outside it, and believe that our ideas about how the universe works are reflections of external reality, rather than the models we have constructed to represent it. We build and modify our internal worlds from the information we detect and the inferences we make, and they are limited because of this. We are unable to conceive of things that do not fit our particular models, so that we can only imagine what a creature from another planet would look like or what happens to us after we die in our own particular frame of reference. The richness of our imagination depends on the richness of our information stores and the ingenuity with which we rearrange the elements they contain. Even the delusions of psychotic individuals are limited by how they understand the world, so that ones who used to believe they were being influenced by a radio transmitter in their brain now believe it is a computer chip.

Predicting the Future

Many of the everyday decisions we make involve predicting the longer-term consequences of the choices that confront us. We base our decisions about getting married, going to college, or investing in the stock market, for instance, on how we predict our choices will turn out. As Holland (1992) notes: "An internal model allows a system to look ahead to the future consequences of current actions, without actually committing itself to those actions. In particular, the system can avoid acts that would set it irretrievably down some road to future disaster." We also use our mental models to predict how external states of affairs, like the economy or the job market, will change over time, either on their own or as the result of some type of intervention. The effectiveness of our predictions depends, of course, on the extent to which our models capture the critical variables and the rules about how they interact. Because these are often difficult to specify with any precision, our models usually contain a number of assumptions, some of which can be quite speculative. As a result, the best that either our brains or our most sophisticated simulations can usually do is to specify a range of possible outcomes and estimate the odds of each of them occurring.[5]

The beliefs and assumptions we develop about the world are incorporated into the models we construct, so that people who have similar beliefs have similar models and see the world in similar ways. These shared belief systems are the glue that binds people together, even when they are based on flawed assumptions. People who believe that taxes should be abolished or deny the Holocaust, for instance, find comfort and validation in being able to connect with each other, even though these ideas may otherwise mislead them. Superstitious beliefs that incorrectly attribute causes also remain with us, since many people still think it is unlucky to walk under a ladder or believe they are in for bad luck if they break a mirror. The models we build can also contain troublesome errors, especially when they are held by political and religious extremists. Some of these individuals are so rigidly certain about their beliefs that they do not hesitate to act against people who hold different ones, rather than try to understand the differences. The better we understand what causes events, however, the more realistic our models, and the more accurate the predictions we base on them.

Truth

We have no way of being certain that the knowledge we accrue is correct, since our information-processing systems are not good truth-detectors. However, this does not stop most people from believing that their ideas and opinions are correct, and that opposing ones are not (Suppe 1977). As Burton (2008) points out, the feeling of *knowing* that we have about the things we believe to be true is an emotion-like state that is not dependent on logical judgment or reason. The strength of someone's convictions is no measure of their accuracy, although it is often mistaken as such. The various ways we have for assessing whether something is true are themselves all subject to error. We can determine, for instance: (a) whether something is inherently plausible (e.g., explainable in natural terms), (b) whether its source is credible (e.g., reliable or authoritative), (c) whether it is consistent with what we already believe, and (d) whether other people whose opinion we value think it is true. Unfortunately, none of these actually *proves* that something is true. In our day-to-day lives, however, we simply assume that information that seems to make sense and fits with our other beliefs is true, and that information that does not is false. Disagreements about who is right and who is wrong about some matter are difficult to resolve, as is evident in the conflicts that arise in the political arena, the courtroom, and domestic disputes. The problem is that the more people's fundamental beliefs are flawed, the more likely they are to believe that information that fits with them is true, even though this only compounds the problem.[6]

Our mental models never fully capture reality, however, because our understanding of the world is built on initial beliefs that themselves cannot be proved. We first have to take something for granted before we can understand something else, an initial point of reference that we have to accept implicitly. Whether we believe that the world is only 5,000 years old, that Fidel Castro planned President Kennedy's assassination, or that the Air Force is suppressing information about flying saucers, depends on the particular models we have constructed, since all of the facts we know are only true within frameworks that may themselves be incorrect. The models we build on faith rather than empiric data, no matter how comforting they may be, are not subject to refutation and are thus of little help in predicting how things will actually turn out. Our senses can also deceive us, as when we see the sun rotate across the sky and think that we are at the center of the universe. Despite these limitations, our species has nevertheless managed to assemble a remarkable edifice of wisdom and knowledge, even if not all of it is universally accepted.

Doubt is the opposite of certainty, and can be just as disabling. Individuals with Obsessive Compulsive Disorder *(maladie du doute)* can be so overwhelmed by it that they cannot make up their mind, do not know what to do, and have to repeatedly check whether they have locked the front door or turned the stove off. A certain amount of doubt is necessary, however, if we want to keep our mind open to the possibility that some seemingly established truths may not be entirely correct. Some people, however, are so uncomfortable tolerating uncertainty that they prematurely rush to closure, even though doing so may preclude them from finding more reasonable alternatives. They make up their mind quickly about matters like evolution or global warming on the basis of what they want to believe, rather than by considering the evidence. Scientists and philosophers, on the other hand, are trained to tolerate a reasonable amount of doubt in order to leave the door open for new discoveries. Although their detractors often take their lack of absolute certainty as a sign of weakness, it is actually an indication of the discipline required to advance our knowledge and understanding of the world (Whitehorn 1963).

Reality

The way meaningful information is detected and processed sheds light on which aspects of the universe exist independently of our awareness of them, and which exist only in the mind of some type of observer. All living creatures are endowed with sensory receptors that enable them to detect and process certain patterns of matter and energy in their surroundings. The particular configurations we are able to detect constitute objective reality for us, since they are the only aspects of the universe that we can sense directly, and the only things we know about that can continue to exist when they are not being detected. The effects these configurations have on us when we bump into them or they bump into us, for instance, also enable to tell that they are there, even if they do not allow us to identify or understand them. Everything else we perceive is subjectively based, shaped in some way by the particular observer.

The observer-dependent features that shape how the world appears to us include all the properties, values, and relationships we attribute to the objects and events we detect. Although we experience these as objective aspects of the external world, this is simply a convenient illusion, since they do not exist independently of the way we perceive them. Just as the smile on the Cheshire cat cannot exist without the cat, properties like big, hot, blue, and the like, cannot exist independent of the objects to which they are attributed and the entities that do the attributing. The only aspects of the universe that our sensory receptors can actually detect are the various chemical and physical patterns they encounter, and our brains then interpret what these signify. We attribute certain properties to recurring aspects of our sensory input, like associating a light wavelength of 6,800 Å with the color red, or labeling sound frequencies above 16,000 Hz as high pitched. We then identify the characteristic features of the various objects and events we encounter by specifying the bundle of properties and relations they appear to possess. A cat, for instance, is a furry, four-legged animal that purrs, likes milk, and says meow, while a dog is one that barks, likes bones, and wags its tail. Our sensory receptors cannot detect a "cat" or a "dog" as such, but only the patterns of matter and energy that conform to how we perceive them.[7]

Although the objective aspects of reality are fixed and nonconditional, the subjective ones are relative to some type of observer-dependent yardstick, based on the particular observer's individual and evolutionary history. Properties like big or little, hot or cold, or sweet or sour are all relative to the particular observer's reference standards and assessments. So also are values like good or bad and right or wrong, as well as spatial and temporal relationships like above or below and fast or slow. Appraisals of whether one object is nearer, colder, or bigger than another one are based on comparisons that occur in the mind of an observer and have no separate existence elsewhere. The reference standard can either be an internal yardstick, another current object, or a designated external benchmark. Assessing an object against an internal standard enables individuals to determine whether it is light or heavy as far as they are concerned, while comparing it with another object lets them establish which is lighter or heavier than the other. Comparison with a consensually established benchmark, however, like a standard meter or a standard kilogram, can determine the actual length or weight of an object in a way that is relatively unaffected by the observer.

Agreeing upon arbitrary reference measures, like the duration of a standard hour or the luminosity of a standard candlepower, enables us to convert certain qualitative assessments into quantitative ones. These external standards permit relatively objective measurements to be made, since they are essentially independent of the person doing the measuring (although they are still subject to observer bias and error). They are not part of external reality, however, since they are based on human-made criteria. There is no natural reason, for instance, for dividing a day into 24 h or a circle into 360°, for any other easily dividable numbers would work just as well. The quantitative measurements that are so essential to scientific research are thus not entirely observer-independent, since they are not inherent properties of the objects and events being measured. While this makes little practical difference within the range in which most measurements are made, it can cause problems at the limits of knowable precision.[8]

Whereas qualitative assessments only determine single, one-at-a-time comparisons, quantitative measurements can be can be repeated by other observers at other times and places and can be compared to, or combined with, other measurements, since they do not depend on the particular observer. Qualitative appraisals are made on an analogic basis by matching a detected information pattern against a reference one, with the output of these comparisons being expressed in relative terms (e.g., more, less, etc.). Quantitative measurements, however, deal with discrete numerical symbols that can be processed digitally, so that they can be combined and manipulated in a variety of ways. While all sentient animals can probably make some types of qualitative assessments, we are the only ones that can make quantitative ones, since these depend on language and conscious reflection. Organisms that do not have any form of consciousness respond to particular configurations of energy and matter on an automatic basis, without attributing any properties or qualities to them—they are just "programmed" to respond to certain detected configurations. A hummingbird does not have to appreciate the red color of a target flower, nor does a female stickleback have to appreciate the red belly color of a potential mate; they just have to have a built-in way of responding to this particular wavelength of light. Fabricated devices, since they also do not have any form of consciousness, also detect and respond to configurations of matter and energy without appreciating what they are detecting.[9]

Notes

[1]Gentner et al. (2001) examine current theories about analogic reasoning and review related research findings. Hofstadter (2007, p. 277) maintains that analogies are a fundamental aspect of human information processing.

[2]Rivett (1972) maintains that: "the history of man is a history of model building, a constant search for pattern and for generalization." He notes that: "a model is first of all a convenient way of representing the total experience we possess, of then deducing from that experience whether we are in the presence of pattern and law and, if so, of showing how such patterns and laws can be used to predict the future."

[3]Universities, encyclopedias, and brains all organize accumulated knowledge into separate sections and departments because the models and assumptions on which the different knowledge sets are based are not readily compatible with each other.

[4]Byrne (2005) believes that engaging in hypothetical thought is one of the major achievements of human cognition. She discusses how people imagine counterfactual alternatives to reality, like thinking about how events might have turned out differently.

[5]Prediction is critical for setting and achieving long-term goals. Comparing predicted outcomes with actual ones provides feedback that enables individuals to modify a related hypothesis or course of action. These types of comparisons are the main way we test our personal, economic, and scientific models (Wiener 1993, p. 118).

[6]Epistemology is the branch of philosophy concerned with the nature of knowledge. Most philosophers define knowledge as justified true belief, implying that one must have a good reason for believing that something is true. But, unfortunately, one person's justification may not be good enough for another. As Franklin (2001) points out, history (what actually happened) and the future (what will happen) do not lend themselves to absolute knowledge of the truth. He describes how the concept of probability evolved as a rational method of dealing with uncertainty in the law and in science.

[7]The various properties we attribute to objects and events generally come in opposing pairs, like fast and slow, high and low, and good and bad (Kelly 1955). The internal reference standards we use to assess them thus appear to involve a continuum that ranges from one extreme of an attribute to its opposite one, rather than being a static representation. We arbitrarily set a mid-point between the two poles of a particular dimension and determine just where particular properties fit on one side or the other of it.

[8]A simple balance scale illustrates how a qualitative assessment can be converted into a quantitative one. For instance: (a) place a stone on one side of the scale and find a second stone that exactly matches its weight, (b) place both of these stones on the same side of the scale and find one that exactly matches their combined weight, (c) transfer this last stone to the pan that already has the two other ones and find one that matches their "4-stone" weight, and (d) repeat the process. Comparable systems have been used to develop the standards we use for measuring the other physical dimensions we employ.

[9]The length of the meter was originally set in 1791 as one ten-millionth of the distance from the earth's equator to the North Pole. This was established in 1889 as being equal to the length of a prototype platinum bar kept in Paris, but was changed in 1983 to being equal to the distance light travels in 1/299,792,458 of a second in a vacuum (since light travels at a speed of 299,792,458 m/s in a vacuum). Even though this new standard is more accurate, it is still an arbitrary, observer-dependent attribute, rather than an entirely objective one.

References

Bruner J (1986) Actual Minds, Possible Worlds. Harvard University Press, Cambridge

Burton RA (2008) On Being Certain: Believing You are Right Even When You Are Not. St. Martin's Press, New York

Byrne RMJ (2005) The Rational Imagination: How People Create Alternatives to Reality. MIT Press, Cambridge

Craik KJW (1943) The Nature of Explanation. Cambridge University Press, Cambridge

Franklin J (2001) The Science of Conjecture: Evidence and Probability before Pascal. Johns Hopkins University Press, Baltimore

Gentner G, Holyoak K, Kokinov B (2001) The Analogical Mind. MIT Press, Cambridge

Hofstadter DR (2007) I Am a Strange Loop. Basic Books, New York

Holland JH (1992) Complex adaptive systems. Daedalus, 121 (Winter): 17–30

Hong FT (2005) A multi-disciplinary survey of biocomputing: 2. Systems and evolutionary levels, and technical applications. In VB Bajic, TT Wee (Eds), Information Processing and Living Systems. Imperial College Press, London

Johnson-Laird PN (1996) Mental Models. Harvard University Press, Cambridge

Johnson-Laird PN (2006) How We Reason. Oxford University Press, New York

Kelly GA (1955) The Psychology of Personal Constructs, Vol 1. WW Norton, New York

Piaget J, Inhelder B (1969) The Psychology of the Child. Trans. H Weaver. Basic Books, New York

Popper KR (1959) The Logic of Scientific Discovery, 2nd edn. Basic Books, New York

Rivett P (1972) Principles of Model Building. John Wiley & Sons, London

Suppe F (1977) The Structure of Scientific Theories. University of Illinois Press, Urbana

Taleb NN (2007) The Black Swan: The Impact of the Highly Improbable. Random House, New York

Whitehorn JC (1963) Education for uncertainty. Perspect Biol Med. 7: 118–123

Wiener N (1993) Invention: The Care and Feeding of Ideas. MIT Press, Cambridge

Chapter 10
Communication

Abstract Living cells and organisms can not only detect meaningful information but can also communicate it to other cells and organisms. Communication is a process in which one entity generates or displays a pattern of matter or energy that is detected by another and results in a change in the latter's behavior, functioning, or structure, without a direct exchange of energy between them. Cues, signals, and messages are the different forms of meaningful information that can be exchanged between biological senders and receivers. The cells of multicellular organisms communicate with one another to coordinate their activities and regulate their functioning. Social animals use their long-distance senses of vision, hearing, and olfaction to transmit information to others and receive it from them. Although energy is required to transmit and respond to information, neither the nature of the information being transmitted nor the response it generates is related to the energy involved.

Cells and organisms are not only able to detect meaningful information but can also communicate it to other cells and organisms. The ability to convey information from one living entity to another is a defining property of biological systems, since the inanimate objects in nature are unable to do this. Communicated information is essential for coordinating the activities of multicellular organisms and for passing genetic instructions from one generation to the next. Information communicated within and between the cells and organs of animals and plants is what enables them to regulate their internal environment and adapt to changes in the external one. Social species also use information conveyed by members of their own kind to regulate their interactions with each other. For meaningful communication to actually take place, however, a receiver must both detect *and* respond in some way to the patterns being transmitted; howling at the moon, signaling for taxis that do not stop, or leaving messages in bottles that do not get found are not here considered to be forms of communication.[1]

A. Reading, *Meaningful Information: The Bridge Between Biology, Brain, and Behavior*,   83
SpringerBriefs in Biology 1, DOI 10.1007/978-1-4614-0158-2_10,
© Springer Science+Business Media, LLC 2011

Communication is a process in which one biological entity generates or displays a pattern of matter or energy that is detected by another and results in a change in the latter's behavior, functioning, or structure, without a direct exchange of energy between them. Cells coordinate their functions by communicating chemically with each other, tissues communicate with other tissues by releasing hormones, sensory receptors communicate with brains, and brains with muscles and glands through neural impulses, members of social species communicate with one another through signals and messages, and generations communicate with succeeding generations through genes and culture. In most of these interactions, information is transmitted as simple on/off signals that activate responses that have been previously programmed into the recipients, either by evolution or learning.

The nature of the information the members of a given cell type or species send and receive has been shaped by natural selection because of the adaptive advantage it has conferred on them (Lewis and Gower 1980). The ability to communicate meaningful information must not only have resulted in an advantage for the senders, it must also have benefited the receivers, for it would not have evolved otherwise (Maynard-Smith and Harper 2003, p. 3). Courtship signals that enhance a male's chances of attracting females willing to mate, for example, increase the likelihood of the male's genes being passed to the next generation, while the ability to detect and respond to these signals also increases the chances of the female's genes being transmitted. The benefits that social communication confer are made clear by the striking structures that several species have developed to facilitate it, such as the peacock's brilliant tail, the elk's ornate antlers, and the bowerbird's elaborate nuptial constructions. They represent expensive investments of these animals' resources that would not have evolved if they did not also confer a significant selective advantage.

Cues, Signals, and Messages

Cues, signals, and messages are the different forms of meaningful information that can be exchanged between biological senders and receivers. Although these are sometimes lumped together under the general rubric of *signs* (i.e., something that *signifies* something), they are treated separately here. The term *cue* is used here to refer to inanimate forms of meaningful information. These can be either naturally occurring cues, like smoke being a sign of fire and leaves changing color a sign of Fall, or ones that are generated inadvertently by living entities, like the footprints Robinson Crusoe saw in the sand or the exhaled carbon dioxide that attracts mosquitoes to mammalian targets. Cues are not considered to be communications, however, since they are not intended to convey information.[2] *Signals,* on the other hand, are patterns of energy or matter that are expressly produced or displayed by one entity in order for them to have an effect on another, like the roar of a lion, the light flashes of a firefly, or human-made traffic signs. *Messages* are complex patterns of *coded* information that can only be responded to by a recipient that can decode them, like the ones conveyed by language and the genome. The lanterns Paul Revere used to warn about the route the British troops were taking were signals ("one if by land, two if by sea"), while the flashing lamp patterns the Navy used to use to communicate between ships at sea were messages.

Communication involves an aspect of the appearance or behavior of one living entity that can be perceived by, and have an effect on, another, like a fox marking its territory with urine deposits or a nightingale singing to attract a mate. As Maynard-Smith and Harper (2003, p. 3) note, signals comprise "any act or structure which alters the behavior of other organisms, which evolved because of that effect, and which is effective because the receiver's response has also evolved." The fact that both the sender and the receiver benefit is one of the other features that distinguish signals from cues. Although cues may benefit the animal that detects them, they are of no advantage to the one that originates them—and may even be a liability. An informational pattern that functions as a signal to one species may serve as a cue to another, like the way the elaborate plumage of certain birds and the loud mating calls of certain frogs make them more noticeable to predators. Although natural selection presumably weeds out cues that are harmful to the animals that emit them, this is offset by how much they benefit them in other ways.

Systems that emit and respond to signals abound in nature, for they are the way that cells communicate with other cells and nonhuman animals communicate with other animals. Signals generally convey simple on/off types of information, so that the responses they generate are determined by the recipients, rather than by the signals themselves. Each species has its own distinctive repertoire of the signals it can send and receive, as well as its own set of functions they serve. Animals send signals to other animals of the same species to coordinate mating rituals, care for their young, control territory, establish dominance, report about food sources, warn of danger, and ensure group cohesion. Howler monkeys, for instance, give warning cries to alert the members of their troop to predators, silverback gorillas beat their chest and rattle nearby bushes to intimidate rivals, and bees dance to tell each other where nectar is located. Signals may also facilitate interaction between differing species, like the way flowers emit scents that attract the insects that pollinate them, possums play dead when confronted by predators, and rattlesnakes use their rattle to warn enemies of their poisonous bite.

Messages are complex patterns of matter or energy that are used to communicate information between individuals of the same species by means of a sequence of coded signals. The two main examples of this are the DNA sequences that convey genetic information and the symbols that comprise human language, both of which are addressed further in later chapters. The responses that messages elicit can vary considerably, since the meaning of spoken and written language is individually interpreted and the meaning of genetic sequences can be affected by contextual factors. Messages differ from signals in that they convey detailed forms of information that can generate responses that are not entirely shaped by the recipients. While signals typically convey only on/off information that activates preprogrammed responses, messages include an informational content that has first to be interpreted and understood, with the way different individuals do this determining how they respond. The complex sequences of sensory signals involved in the notes of a bird's song and in the back and forth communication that occurs during many courtship rituals do not constitute messages, however, since they follow fixed patterns and elicit fixed responses.

Animal Communication

Animals generally use their distance senses of vision, hearing, and olfaction to transmit information to others and to receive it from them, although some also communicate directly through touch. Visual information is usually conveyed through bodily appearance or behavior, like the elaborate plumages and mating rituals of certain birds, while auditory information is transmitted through sounds and vocalizations, like the bark of a dog or the buzzing of a cicada. As Deacon (1997, p. 31) notes: "Many birds, primates, and social carnivores use extensive vocal and gestural repertoires to structure their large social groups. These provide a medium for organizing foraging and group movement, for identifying individuals, for maintaining and restructuring multidimensional social hierarchies, for mediating aggressive encounters, for soliciting aid, and for warning of dangers." Animals are also able to detect meaningful cues about members of their own or other species, such as where they are and what they are doing. Birds flying in a flock or fish swimming in a school, for instance, are not necessarily communicating with each other, although they are perceiving cues about each other's position and movement (Evans 1968).

Most animal communication triggers innately programmed behaviors in which a signal from one animal elicits a fixed response in another. Many species of birds and fish perform ritualized "dances" in which a brightly colored male engages in a series of stereotyped movements and/or vocalizations that attract a responsive female and culminate in fertilization and egg laying. The bright red underbelly that male three-spiked sticklebacks develop in response to lengthening daylight attracts conspecific females; and the male then swims around the female in a zigzag pattern that induces her to follow him and spawn in the nest he has prepared. Experiments with various models have shown that it is the red color that elicits the female's response—which was also produced by red postal vans passing by the window of the English laboratory where the fish were housed (Tinbergen 1953). Courtship behaviors can serve other adaptive functions in addition to facilitating mating, such as inducing hormonal changes that stimulate ovulation or functioning as a barrier to interspecies breeding. Different species of ducks, for instance, make distinctive types of head movements during courtship, but respond only to those of their own particular kind (Judson 2002).

Animals can communicate meaningful information to each other about a variety of current matters, such as their sexual readiness or aggressive intent, but not about past or future concerns. Darwin (1872) noted that social animals often communicate information about their emotional state, like the way a dog wags its tail when happy, tucks it between its legs when fearful, and holds it rigidly erect when hostile. While we can usually infer the meaning of the information animals convey to each other from the effect it has on the recipients, we do not have any direct way of knowing what they actually make of it. Vervet monkeys, for example, use three acoustically distinct alarm calls to alert their troop to different types of predators, with each of them eliciting a different adaptive response: the "leopard alarm" gets them to run to the nearby trees, the "eagle alarm" to run into the surrounding bushes, and the "snake alarm" to move to a safe distance from the source (Cheney and Seyfarth 1990). Many species also communicate through distinctive threat and dominance behaviors, such as those displayed by males during competition over food, mates, or territory. Dogs bite at each other, robins display their red breasts, wolves growl and bare their teeth, bull elephants trumpet, and chickens peck each other, all for the sake of letting others know of their presence and intentions (Alcock 2005).

Transmission

Energy is required to transmit meaningful information to a distant recipient, but neither the nature of the information being transmitted nor the response it generates is related to the energy used to transmit it. Although information requires a vehicle of some sort for its transmission, the vehicle is not the message. It is the *pattern* of energy or matter that conveys the meaning, which is why the same amount of energy can transmit information that has a completely different meaning, as in "the boy hit the ball" and "the ball hit the boy." The transmitting and receiving entities both have to have their own source of energy in biological communication systems, just as they do in electronic ones.[3]

Even though energy is needed to transmit signals between senders and receivers, nothing material is conveyed between them. Biological information is always transmitted in a one-way direction that cannot be reversed, since biological signals only trigger the receiver's response, but do not determine it the way television signals do. The television shows we watch are not preprogrammed into our sets and then just activated by signals from the stations, for the stations transmit their content, not just on/off signals. The irreversibility of genetically transmitted information is the basis of what Crick (1970) calls "the central dogma of molecular biology"—the principle that information cannot be transferred back from proteins to the genome.

Notes

[1] The field of *Biosemiotics* developed by Thomas Sebeok (1920–2001) involves the study of communicative signs in living systems. *Semiotics*, the discipline from which it is derived, is concerned primarily with understanding *signification*, rather than *information*. It regards all information as *signs*, which are conceptualized along lines pioneered by C. S. Peirce (1839–1914) as things that stand for something else, to someone, in some capacity. Biosemiotics applies concepts developed initially for understanding human communication to a range of biological phenomena, including genetic and cellular processes. However, because of its different orientation, the way it

deals with these topics differs from the one proposed here. Information about Biosemiotics can be found in Sebeok and Umiker-Sebeok (1992), Petrilli and Ponzio (2001), Barbieri (2003), and Barbieri (2006).

²The idea that signals and messages are different from cues because they (signals and messages) are *intentionally* generated is difficult to apply to entities that do not have conscious intentions. The approach taken here is that signals in organisms that lack conscious choice have been shaped by natural selection and are generated automatically. The meaning of cues, like storm clouds in the sky or brown patches in a lawn, is generally learned from the effects they produce.

³Hieroglyphics, notched sticks, and cave art were probably the first relatively lasting ways of transmitting information to recipients who were not physically present. The first commercial system for sending information over distances was the electric telegraph that Samuel Morse and Alfred Vail patented in 1837. Its messages were transmitted in "Morse Code," a system that represents the alphabet by dots and dashes that are conveyed by electric currents of either short or long duration. The first long-distance voice transmission was sent over a copper wire between Washington and Baltimore in 1844, which was followed by Alexander Graham Bell's introduction of the telephone in 1876. Wireless forms of transmission were then developed by Guglielmo Marconi in the late 1890s (Pierce 1980).

References

Alcock J (2005) Animal Behavior: An Evolutionary Approach, 8th edn. Sinauer Associates, Sunderland

Barbieri M (2003) The Organic Codes: an Introduction to Semantic Biology. Cambridge University Press, Cambridge

Barbieri M (2006) Introduction to Biosemiotics: The New Biological Synthesis. Springer, Dordrecht

Cheney DL, Seyfarth RM (1990) How Monkeys See the World. University of Chicago Press, Chicago

Crick F (1970) Central dogma of molecular biology. Nature 227: 561–563

Deacon TW (1997) The Symbolic Species: The Co-Evolution of Language and the Brain. WW Norton, New York

Darwin C (1872, 1934) The Expression of the Emotions in Man and Animals. Abridged edn. Watts, London

Evans WF (1968) Communication in the Animal World. Crowell, New York

Judson O (2002) Dr. Tatiana's Sex Advice to All Creation: The Definitive Guide to the Evolutionary Biology of Sex. Metropolitan Books, New York

Lewis BD, Gower DM (1980) Biology of Communication. John Wiley & Sons, New York

Maynard-Smith J, Harper D (2003) Animal Signals. Oxford University Press, Oxford

Petrilli S, Ponzio A (2001) Thomas Sebeok and the Signs of Life. Icon Books, Cambridge

Pierce JR (1980) An Introduction to Information Theory: Symbols, Signals and Noise, 2nd edn. Dover Publications, New York

Sebeok TA, Umiker-Sebeok J (1992) Biosemiotics. The Semiotic Web. Mouton de Gruyter, New York

Tinbergen N (1953) Social Behavior in Animals: With Special Reference to Vertebrates. John Wiley & Sons, New York

Chapter 11
Language

Abstract A language is a system of grammatically organized symbols that enables us to communicate meaningful information to others. Words are arbitrary symbols that refer to objects and events in the natural world, as well as to intangible ideas and relationships derived from them. While signals are limited to providing on/off information, language-based messages are able to convey abstract thoughts, tell stories, ask questions, and transmit acquired knowledge. Meaningful discourse involves language that not only is intended to have an effect on one or more recipients, but also does have an effect on them, even if it is not the one intended. We also communicate meaningful information nonverbally, both in association with language and apart from it. Nonverbal messages only convey information about the sender's current emotional state or behavior, which is why animals and infants can only communicate how they feel or what they want, not what they think or know.

Language is a unique medium that we have developed for conveying, storing, and manipulating complex forms of meaningful information. It is probably the attribute that most distinguishes us from the rest of the other species, for, without it, we would probably be just another collection of nomadic primates roaming the subtropics in search of food and shelter. Symbolic language is a relatively recent development in our evolutionary journey, most likely dating back just over 45,000 years to the time when other forms of symbolic activity first appeared, like cave paintings and burial rituals (Deacon 1997). Written languages emerged about 6,000 years ago, although they did not become widespread until the invention of the printing press some 5,500 years later. Our ancestors previously depended on oral stories to pass acquired information on to succeeding generations, but printing allowed them to store and disseminate it on a much wider scale. The development of the Internet is once again transforming society by making access to stored information almost limitless, even though it may not all be meaningful.

A. Reading, *Meaningful Information: The Bridge Between Biology, Brain, and Behavior*, 89
SpringerBriefs in Biology 1, DOI 10.1007/978-1-4614-0158-2_11,
© Springer Science+Business Media, LLC 2011

A language is a grammatically organized system of symbols that is used to communicate meaningful information to others, either by means of audible sounds (speech), visual images (writing), observable gestures (signing), or tactile stimuli (Braille). There are over 5,000 known languages, most of which have their own set of symbols, their own code for determining what they represent, and their own rules for combining them into intelligible messages (Diamond 1997). Both sender and receiver have to understand the code and the rules being used in order for meaningful information to be communicated, since detected linguistic symbols are otherwise meaningless. Language's ability to transmit acquired information has led to a steady increase in our knowledge and understanding of the world and allowed us to develop increasingly flexible ways of adapting to it. Each of us inherits a legacy of culturally transmitted information that allows us to stand on the intellectual shoulders of our predecessors. Other species, however, are left to cope with the world largely in terms of their own experience, so that each generation has to begin anew, unable to benefit from what others have previously learned.[1]

Words

Language conveys information mainly by means of spoken and written words. The letters of the alphabet are agreed-upon artifacts that can be combined to form words, which can then be linked together to create phrases, sentences, and messages. Words are arbitrary symbols that not only refer to objects and events in the natural world, but also to intangible ideas and relationships that have been derived from them. Language's ability to represent facts and abstract concepts that are not directly perceivable greatly expands the range of information it can convey. Words are symbols that we respond to in much the same way we respond to the entities they represent. But, unlike the objects they represent, words are digital entities that can be combined and rearranged to express complex thoughts, concoct stories, and construct mental models.

Written messages are generally comprised of a series of sentences, each of which contains various parts of speech that are linked together by a set of grammatical rules (*syntax*). The different parts of speech have different functions. Nouns are used to represent objects; verbs to represent actions and relationships; and adjectives, adverbs, participles, and articles to modify the meaning of nouns and verbs. These lexical elements enable us to compress the information we store, since we do not have to save every message we receive in its entirety. The same adjective can be used to modify any number of nouns without each particular combination having to be stored, and verbs can specify the relationship between different entities without being locked into any of them. This is a highly efficient way of communicating and storing information, since symbols can be combined to portray an endless assortment of objects and events, as well as their properties and relationships (Bickerton 1990).

Individual words, isolated from any context, do not ordinarily convey meaningful information, even though they still have a dictionary-type meaning that signifies the object, event, or relationship they represent. This literal meaning does not by itself generate a response in the recipient—any more than the object it represents generates one by itself, apart from the context in which it is perceived. What a word means simply denotes what it refers to, not the effect it can generate as part of a message. The dictionary meaning of the word "snow" (a frozen, crystalline state of water that falls as precipitation) is thus different from its informational meaning, since the latter depends on the effect it has on particular recipients. While the meaning of words that refer to tangible objects can generally be established by consensus, reaching agreement about the meaning of ones that refer to abstract entities is not always easy, so that words like *democracy* and *socialism* can mean different things to different people. The context is always critical in communicating meaningful information, for the same word can refer to a number of different things, like the *bark* of a dog or the *bark* of a tree. Although a few of them are able to function as meaningful signals on their own (e.g., *hello, yes, stop*), words ordinarily communicate meaningful information only when combined with other words to form sentences and messages.

Messages

One of the unique features of language is the way messages can convey complex forms of information, not just on/off signals. Messages are composed of a series of words strung together according to prespecified rules, which allow them to convey abstract thoughts, tell stories, ask questions, and transmit acquired knowledge. Messages have to be decoded, however, to learn their literal meaning, and then interpreted to discover their actual one—since what words *denote* (their literal meaning) and what they *connote* (their implied meaning) are not always the same. The implied meaning of a spoken or written message is determined by a number of factors, including the relationship between the sender and the receiver, their associations to the words being used, and the linguistic and situational contexts in which it is being transmitted. Although the words in a message have predetermined (literal) meanings, the information they convey does not, since its meaning is determined by the effect it has on the recipient.[2]

Messages do not trigger preprogrammed responses the way signals do, for the effects they generate depend on how the recipients understand and interpret them. No matter what the intent of the senders, messages only convey meaningful information to recipients in whom they generate a behavioral, physiological, or neuronal response. Messages that are not conveyed intentionally, such as Freudian slips and overheard conversations, are not considered as communications, even when they have an effect on the recipients. People can also reveal information inadvertently about their level of education, social class, and region of origin through the choice of words and idioms they use and the way they respond to certain topics. Even things that are not said can convey meaningful information, like failing to remember an anniversary or answering a question evasively. Meaningful information that is not meant to be communicated to the recipient who detects it is, however, considered to be a cue, not a message.

Meaningful Discourse

Meaningful discourse involves the expression of language that is not only intended to have an effect on one or more recipients, but actually does have an effect on them, even if it is not the one the sender intended. Messages that convey meaningful information are generally designed to achieve one of three goals: either to induce other individuals to do something, feel something, or know something, that is, to generate a change in their behavior, physiologic state, or neuronal connections. Novels, plays, television shows, movies, concerts, magazines, scientific reports, news media, advertisements, and graffiti can all be forms of meaningful discourse as long as the information they convey causes at least some individuals to do something, feel something, or know something that they otherwise would not have.

Language is the only form of communication in which the information being transmitted has to have a meaning to the sender as well as the receiver. The meaning that messages have for the senders is determined by the effects they intend to have on the recipients, while the effects they actually elicit represent the meaning they have for the recipients. Problems arise, however, when a recipient perceives a meaning that is different from the one the sender intended, since effective communication requires that both understand it the same way. Otherwise misunderstandings occur. Misunderstood information may still be meaningful, however, as long as it generates a response in the recipient. People may, for example, take offense at a remark when none is intended or mistakenly assume that another person dislikes them and respond accordingly. Miscommunication can even occur when the sender and the receiver both understand the literal meaning of the words involved, because languages use a variety of nonliteral "figures of speech" to add emphasis or color to the ideas being expressed. These include metaphors (*food for thought*), idioms (*eat your heart out*), slang (*neat* and *cool*), technical jargon (*free-electron mass*), euphemisms (*revenue enhancement*), and figurative expressions (*couch potato*) (Pinker 1997).[3]

Spoken language usually conveys relationship information as well as content information (i.e., a meta-message as well as a message). Senders choose the way they express what they want to say in part to convey their perceived or desired relationship with the receiver, in terms of status, intimacy, respect, and so on (Watzlawick et al. 1967). As already mentioned, information does not have to be true to be meaningful. People believe and respond to an array of claims for miracle cures, implausible rumors, and far-fetched schemes, even when the information involved seems too good to be true. Although language is humanity's most valued servant, it can be a treacherous one at times, since prejudices, false beliefs, and disinformation can be transmitted by it just as easily as information that is fair and factual—and can just as easily be incorporated into a recipient's worldview.

Nonverbal Communication

Humans can also communicate meaningful information to each other by nonverbal means, both associated with language and separate from it. The meaning of spoken messages can be affected by the nonverbal expressions that accompany them, including the sender's emphasis, tone of voice, facial expression, gestures, and body language. Although there is generally a congruence between the verbal and nonverbal information being conveyed, there can be discrepancies between them, like expressing affection verbally while remaining emotionally distant. Nonverbal information can also be communicated apart from language, either as sounds like laughing or crying, visual displays like grimacing or yawning, or tactile contact like hitting or kissing. Nonverbal communication is not able to convey information about the objective features of the world, since it is limited to expressing data about the sender's emotional or behavioral state. This is why animals and infants can only communicate how they feel or what they want, not what they know. Gestures and vocalizations are only able to refer to things in the sender's current environment; they cannot communicate abstract ideas or refer to past or future events (Wolfgang 1979).[4]

All of us still retain some of the nonverbal behaviors our ancestors must have used before they acquired symbolic language. These include the expressions of emotion that intrigued Darwin—the signs of anger, fear, and joy that can be recognized on our faces. The fact that every culture expresses basic emotions like happiness, sadness, anger, and fear in the same way points toward their evolutionary origin and explains why they are usually generated and responded to automatically. Humans have also developed culturally derived forms of nonverbal communication, including music, art, and traffic lights. Much of what we do, how we dress, what we drive, and where we live conveys nonverbal information to others, either intentionally or otherwise. Although a number of other species have developed fairly elaborate systems of nonverbal communication, their vocalizations and gestures are not genuine languages, nor are the hand signs that domesticated chimpanzees have been taught to use, since they lack the structure and flexibility of true sentences (Greenspan and Shanker 2004).

Neural Representation

No one knows exactly how language is represented in the brain, how we find the right words to express what we want to say, or how we decode the symbols others convey to us. Although our conscious thoughts generally take the form of spoken language, we do not know how the brain processes the information they contain, since this takes place outside of our awareness. The way we are able to conjure up novel stories in our imagination that are not like anything we've ever experienced implies, however, that the elements from which these are constructed are stored as discrete, combinable entities (Fauconnier and Turner 2002). But, while we can envisage ways that individual words might be represented by specific patterns of neural connectivity, we have no idea about how the brain actually links them together to form intelligible messages (Pinker 2007).

The brain apparently processes spoken and written language symbols by combining digital and analog types of information processing. The meaning of individual words (i.e., what they denote) can be ascertained by matching them against the stored reference patterns that represent the particular individual's vocabulary. But this type of analogic matching cannot be done with messages since they do not have preexisting reference patterns. The meaning of words that are linked together in a message (i.e., what they connote) is an emergent property, since it not only depends on the meaning of the individual words, but also on the way they are organized and arranged. Symbolic information offers an immense advantage over the sensory types of information we share with other animals, as the latter simply records events without being able to analyze, integrate, or reconstruct them. Because they are not able to process symbolic language, most animals live their days in the present, reacting only to the events they experience at the particular moment or to the associations these elicit. As McCrone (1991) observes: "We would find animal minds strangely uncluttered. There would not be the same churning of past thoughts and future plans that fill the human mind. There would not be the continuous chatter of our inner voice, not the sudden breaks to reconsider our own actions as we switch from simple awareness to self-awareness."

Notes

[1] There are a number of books about the nature of language, including Bickerton (1990), Pinker (1994), Bickerton (1995), Deacon (1997), Pinker (1999), and Jackendorf (2002), although none of them deals with the way it conveys meaningful information.

[2] *Semantics* is the study of meaning in language. It focuses on the literal (dictionary) meaning of words, and then uses these to understand the meaning of the phrases and sentences formed from them, separate from any effect these have on the recipients. Most linguists and philosophers focus more on the meaning of words than the meaning of the information being conveyed by them (Putnam 1975). As Johnson-Laird (1996, p. 182) observes: "the true nature of meaning has puzzled thinkers from the very beginning of philosophy, since they can't agree whether meanings are aspects of the natural world or mental constructions imposed upon it."

[3] According to Boroditsky (2009), the different languages we speak also "shape the way we think, the way we see the world, the way we live our lives" because of the different ways they conceptualize abstract phenomena, such as space, time, and relationships.

[4] The now-discredited "double-bind" theory of schizophrenia was based on an assumed dissonance between the verbal and nonverbal messages a mother supposedly conveyed to her child (Bateson et al. 1956). Conflicts between verbal and nonverbal messages are, however, still a significant source of misunderstanding and disagreement, especially between individuals who are emotionally involved with each other.

References

Bateson G, Jackson DD, Haley J, Weakland J (1956) Toward a theory of schizophrenia. Behav Sci 11: 251–264

Bickerton D (1990) Language & Species. University of Chicago Press, Chicago

Bickerton D (1995) Language and Human Behavior. University of Washington Press, Seattle

Boroditsky L (2009). How does our language shape the way we think? In M Brockman (Ed), What's Next: Dispatches on the Future of Science. Vintage Books New York

Deacon TW (1997) The Symbolic Species: The Co-Evolution of Language and the Brain. WW Norton, New York

Diamond J (1997) Guns, Germs, and Steel. WW Norton, New York

Fauconnier G, Turner M (2002) The Way We Think: Conceptual Blending and the Mind's Hidden Complexities. Basic Books, New York

Greenspan SI, Shanker SG (2004) The First Idea: How Symbols, Language, and Intelligence Evolved from Our Primate Ancestors and Modern Humans. Da Capo Press, Cambridge

Jackendorf R (2002) Foundations of Language: Brain, Meaning, Grammar, Evolution. Oxford University Press, New York

Johnson-Laird PN (1996) Mental Models. Harvard University Press, Cambridge

McCrone J (1991) The Ape That Spoke. William Morrow, New York

Pinker S (1994) The Language Instinct. William Morrow, New York

Pinker S (1997) How the Mind Works. WW Norton, New York

Pinker S (1999) Words and Rules: The Ingredients of Language. Basic Books, New York

Pinker S (2007) The Stuff of Thought: Language as a Window into Human Nature. Viking, New York

Putnam H (1975) Mind, Language and Reality. Philosophical Papers Vol. 2. Cambridge University Press, Cambridge

Watzlawick P, Beavin JH, Jackson DD (1967) Pragmatics of Human Communication: A Study of Interactional Patterns, Pathologies, and Paradoxes. W.W. Norton, New York

Wolfgang A (1979) Nonverbal Behavior: Applications and Cultural Implications. Academic Press, New York

Chapter 12
Cellular Signals

Abstract Cells are the basic building blocks from which all of the dazzling array of nature's creatures have been fashioned. The primary way that cells communicate with each other is by releasing various signaling molecules into their surroundings, where they bind to and activate specific protein receptors on other cells. Proteins are extremely large biological molecules that can detect and respond to informational signals. They regulate the cell's metabolic activities by turning on chemical reactions that are currently needed and turning off ones that are not. Although each cell contains an organism's entire genome, only the genes specifying the proteins it needs to perform its current function are activated. Neurons are cells that specialize in transmitting signals from one part of the body to another and forming patterns of connectivity that allow the brain to store and represent information.

Cells are the basic units of all living things, the tiny metabolic factories that make life possible. They are the building blocks from which the dazzling array of nature's creatures has been fashioned, as well as the infinite variety of their shapes, sizes, constituents, and behaviors. A multicellular organism is like an immense corporation that is organized into a number of specialized departments (the organs) and manned by a multitude of dedicated workers (the cells), each of which is either taking in supplies, manufacturing substances, repairing damaged equipment, getting rid of waste, or communicating with others—all in strict accord with some type of built-in master plan. The inside of a cell is a tiny chemical world, packed with a myriad of different types of molecules whose interactions with each other are regulated by a network of protein-based signaling mechanisms. Cells coordinate their activities by sending and receiving chemical signals that regulate their internal milieu and enable them to adapt to their external environment (Niehoff 2005).[1]

Most cells consist of three basic components: a nucleus that contains their chromosomes and genes, the cytoplasm, a surrounding "soup" that contains several thousand chemical substances, and an enclosing membrane that separates the cell

A. Reading, *Meaningful Information: The Bridge Between Biology, Brain, and Behavior*, SpringerBriefs in Biology 1, DOI 10.1007/978-1-4614-0158-2_12, © Springer Science+Business Media, LLC 2011

from its immediate surroundings. Virtually every cell in an organism contains a complete set of its genes, but only those related to a particular cell's functions are expressed (activated) in it. A liver cell differs from a kidney or skin cell, for instance, by the set of genes that can be turned on in it. Stem cells, from which all the other cell types are derived during development, are the only ones that can have all of their genes expressed, which is why there is such an interest in seeing whether they can be used to repair damaged tissues. The genes that are expressed in a given cell determine which proteins the cell produces, and these then regulate the cell's growth and metabolic functions.[2]

Considering that typical vertebrate cells are so small that they can only be seen with a microscope, the complexity of the interactions that take place within them is truly amazing. The genes in the cell's nucleus regulate the chemical reactions in its cytoplasm by turning on those that are currently needed and turning off those that are not. They do this by specifying the types of proteins the cell makes, which is what determines which reactions take place. This process ordinarily triggers just the right chemical interactions, in just the right amounts, in just the right cells, at just the right time, to enable animals to convert food into energy, maintain homeostasis, swim, walk or fly, reproduce, and respond to incoming information. The ability to detect and respond to meaningful information conveyed both between cells and within them is essential for sustaining the health and wellbeing of all living things. Errors in the way cells process information result in the types of abnormal molecular functioning that characterize diseases like cancer and diabetes.

Cellular Communication

The primary way that cells communicate with each other is by releasing various signaling molecules (called *ligands*) into their surroundings where they bind to and activate specific protein receptors on other cells. These receptor proteins are generally embedded in the membrane around the receiving cell in such a way that part of their structure extends outside the cell, where it can make contact with its ligand, and part intrudes into the cell, where it can interact with intracellular chemicals. The receptor proteins change their three-dimensional shape when their matching ligand binds to them, and this then triggers a series of reactions inside the cell. Ligand molecules can either be transmitted by diffusion to nearby cells or secreted into the circulation as hormones that act on more distant ones. Although hormones are the best-known group of signaling molecules, there are a great many others, including growth factors, cytokines, and neurotransmitters. Multicellular animals coordinate the biochemical reactions that take place within their various tissues and regulate their interactions with the surrounding world through a constant chatter of such ligand-based signals.[3]

Receptor proteins typically respond only to the specific ligand for which they have affinity, so that molecules that differ just slightly from their particular ligand tend to have no effect on them. Ligands that activate receptors are called agonists, while those that bind to receptors without activating them are called antagonists, because they block the agonists from having an effect. The endorphins, for example, are naturally occurring ligands that activate certain receptor proteins in the brain, thereby causing them to release dopamine and other chemicals that produce a highly pleasurable state in the individual. Exogenous substances like morphine and heroin, because they activate the same receptors, produce the same pleasurable effects. These types of binding affinities play a large role in the development of new pharmaceutical substances, since many medications are made of small organic molecules that act either as agonists or antagonists on various cell receptors. Some of these substances are thus able to counteract the effects of certain disorders by either activating or inhibiting particular cellular functions.

Membrane-bound proteins respond in highly specific ways when a signal molecule activates them. Some react locally, such as by opening up a channel in the membrane that allows certain charged ions to enter or exit the cell, while others initiate a series of chemical and informational activities inside the cell. The latter are referred to as *second messengers* because they augment the effects of the initial signal by launching a cascade of intracellular ones that catalyze particular reactions and/or switch on or off various genes. The aggregate of all the cellular interactions taking place within an organism determines its physiological and behavioral activity at that time.

Proteins

Proteins are extremely large molecules made up of several hundred amino acids arranged in sequence along a linear chain. Twenty different amino acids are used in making proteins, and the particular ones involved and the particular order in which they are arranged provide each protein with a distinctive identity. What makes proteins so special is how their long molecules fold on themselves in a way that gives each of them a unique three-dimensional configuration that has nooks and crannies into which small molecules (i.e., ligands) can fit. The mechanism is highly selective, in that each protein tends to bond only with a specific ligand and, when it does, this alters the protein's shape in a way that activates a particular cellular response. The proteins respond to the spatial configuration of their ligand, not to its chemistry, and the response this sets off is energized by other substances in the cell, not by the ligand or the protein. The process is thus primarily an informational one, not a physical one, as Monod (1971, p. 78) notes: "The huge network of cybernetic interconnections makes each organism an autonomous functional unit whose performances appear to transcend the laws of chemistry, if not to ignore them altogether."[4]

Proteins are truly unique molecules, because they are the only biological substances that can detect and respond directly to meaningful information. They are, in fact, nature's fundamental information-detecting devices, without which life as we know it would not be possible. There are a number of proteins, however, that have nothing to do with information, but are involved instead in structural functions, like the way the long filaments of *actin* and *myosin* support muscle contraction.

Although living entities are essentially made from the same basic chemicals as nonliving ones, they are unique in the way these are arranged into highly organized structures, like proteins, cells, and tissues, and in the way these communicate with each other. Protein macromolecules are the signaling system that orchestrates this self-organizing process by telling each cell exactly what to do and when to do it, during both the development and lifetime of every living organism. As Küppers (1990, p. xviii) notes: "The material order and the purposiveness characteristic of living systems are governed completely by information, which in turn has its foundation at the level of biological macromolecules."

The distinctive spatial configurations of protein molecules enable them to detect and respond to the chemical signals that convey meaningful information within and between cells. It has been estimated that human tissues contain about 100,000 different proteins, although not all of them are present in every cell. Each cell type contains its own particular complement of several thousand of them, with particular ones being activated as needed to carry out its functions. Which chemical reactions take place in a given cell, when they take place, and for how long is controlled by informational signals that are detected and responded to by proteins. The way the cellular components work together to perform their designated functions may be nothing short of a miracle, but it is a miracle orchestrated by the way that proteins detect and respond to meaningful information.

Some of the proteins inside cells function as *enzymes*, substances that, when activated, dramatically accelerate the rate of particular chemical reactions. Enzymes enable certain metabolic activities to proceed thousands of times faster than they would have done otherwise, so that organisms can respond rapidly to changing circumstances. Most of the enzymes in a cell are highly selective, since they are only activated by specific signaling molecules and only catalyze specific chemical reactions. Enzymes do not participate directly in the chemical reactions they catalyze, nor do they provide the energy needed to sustain them, as other substances in the cell supply this. They function as information detectors that respond to particular ligands by changing their shape in ways that initiate their particular catalytic processes. The set of enzymes that are turned on in a cell at any given time determines the metabolic pathways that are currently active in it, as well as the physiological functions these facilitate (Cornish-Bowden 2004).

This process by which an extracellular signal molecule activates a series of intracellular proteins is usually referred to as *signal transduction* because of the way it converts one kind of signal into another. The chain of signals is like a sequence of relay stations, each of which is activated by a particular molecular signal and responds by emitting one of its own, until the final outcome is achieved. Because these chemical reactions and informational signals occur at microscopic and millisecond levels, it is not easy to tease them apart, although there are important differences between them. Informational transactions differ from chemical ones because both the effects they cause and the energy needed to produce them are functions of the recipient molecules, not the initiating ones, and thus cannot be predicted by ordinary physical laws. Like the appliances in our homes, the metabolic processes in our cells have to have an independent supply of energy in order to function when they are switched on.[5]

Animals obtain the energy they need to maintain their current degree of organization (and to increase it during periods of growth and development) from the foods they ingest, since these release energy as they get digested into simpler molecules. The amount of energy that foodstuffs release when they are broken down into their constituents is measured in calories. If the amount of calories produced by an animal's food intake exceeds its energy requirements, the excess is stored as fat, which can be mobilized whenever additional energy supplies are needed. There are two basic types of metabolic activities in animal cells: *catabolism*, in which complex molecules are broken down into less complex ones in order to produce available energy, and *anabolism*, in which available energy is used to build complex substances from simpler molecules. Carbohydrates, for instance, release energy when they are oxidized in the body to carbon dioxide and water, and this energy is then captured by adenosine diphosphate molecules (ADP), which are thereby converted to adenosine triphosphate molecules (ATP). The ATP molecules then donate this energy to various cellular processes by converting back to ADP, with the two of them going through the same routine over and over again, much like repeatedly recharging a battery.

Neurons

Neurons are specialized cells that communicate information rapidly from one part of the body to another. They are made up of three components: a cell body, which contains its nucleus and genetic material, a bushy network of protoplasmic filaments (the *dendrites*) that protrudes from one side of the cell body and contains up to several thousand receptors, and a single, long filament (the *axon*) that extends out from the opposite side. Incoming information is conveyed to the dendrites by chemical signals, and outgoing information is transmitted along the axon by electric pulses, called *action potentials*. Neurons communicate with one another through a series of *synapses*, minute gaps between them where the axon of one lies in close proximity to a dendrite of another so that chemical signals can readily pass between them. Each neuron has a string of outgoing synaptic terminals along the length of its axon, all of which contain microscopic sacs of neurotransmitter chemicals that are released into the adjacent synapses when an action potential travels down the axon. These neurotransmitters then bind with receptor proteins on the dendrites of adjacent neurons, which changes the three-dimensional shape of the receptors in ways that alter the electrical potential across the cell membrane (Cairns-Smith 1996). Excitatory impulses let positively charged ions into the cell, while inhibitory ones let negatively charged ones in, with an action potential being generated in the receiving neuron when the net amount of positive charge produced by incoming signals reaches a threshold value. The electrical impulse that this initiates travels down the axon and triggers the release of transmitter chemicals into the adjacent set of synapses, where the signal is passed on to the nearby neurons. The whole cycle takes only a few milliseconds and then, after a brief pause, it is ready to begin again (Stevens 1995).

The action potentials that different neurons generate are virtually indistinguishable from one another, and thus carry no information about the nature of the stimulus that generated them. They are, however, able to convey other types of information: (a) the rate of firing of the impulses conveys information about the intensity of the stimulus, (b) the neural pathway along which its signals travel convey information about the site and nature of the originating stimulus, and (c) the pattern of activity in a group of related neurons conveys information about the pattern of the originating stimulus (as in vision and hearing). The information that is communicated to the brain is thus not the same as the information the sensory receptors detect; it is just a set of signals that indicate that a particular configuration of matter or energy has been detected (Whitfield 1984, p. 133). The size or form of the action potentials generated by a given neuron are independent of the nature of the signals it receives; a neuron either generates its characteristic axon spike or it does not, since this is an all-or-none process. This makes it possible, however, for the array of on/off states in a set of neurons to act as a coding system that can transmit digital types of information (Koch 1999, p. 142).

The human brain contains about 100,000,000,000 neurons, each with an average of about 8,000–10,000 incoming synaptic connections on its dendrites and an equal number of outgoing ones on its axon (Black 1991). The intricate way that neurons network with each other and the endless patterns of connectivity they can form are what make it possible for highly complex information patterns to be represented in the brain. As (Glynn 1999, p. 140) points out: "The energy of the nerve impulse comes from metabolic processes within the nerve fibre, and what is significant is not the transfer of energy—which at best is tiny and may be nonexistent—but the transfer of information." The majority of the neurons in the adult brain do not undergo cell division the way that other cells do, so they cannot be replaced when lost through injury or disease. Neurons are unlike other cells in that they are affected by what they experience, as shown by the way their synaptic connections change in response to the detection of meaningful information. The learning and memories that neurons enable animals to develop are incorporated into their network of connections, not in the neurons themselves, and thus cannot be transferred to newly formed neurons. New neurons are, nevertheless, formed in a few specific areas of the brain, such as the hippocampus, but their function is not entirely clear.[6]

Notes

[1] Weisskopf (1966), Dulbecco (1987), Loewenstein (1999), Harold (2001), and Noble (2006) provide overviews of molecular and cellular biology.

[2] The two fundamental domains of living organisms are the Eukaryotes (multicellular animals) and the Prokaryotes, with the latter being comprised of the Eubacteria (bacteria) and the Archaea (methane producing microorganisms that exist in inhospitable locations). The Eukaryotes are the only organisms whose cells contain an identifiable nucleus, although vertebrate germ cells and red blood cells do not have one.

[3] A few ligands actually pass through the cell membrane and act directly on protein receptors inside the cell, including thyroid and steroid hormones. Some cell-to-cell communication also involves direct contact through the formation of *gap junctions* that connect the cytoplasm of one cell directly

to the cytoplasm of an adjacent one. Gap junctions play an important role in cardiac muscle, where they allow signals from the pacemaker region to spread rapidly between adjacent cells in order to coordinate the heart's contractions (Holcombe and Paton 1998).

[4] *Proteomics* is the study of the structure and function of proteins as vital constituents of living organisms. The *proteome* is the set of proteins produced by an organism during its lifetime, in the same way that its genome is its set of genes. While the genome is virtually identical in every cell of a given organism, the proteome varies from one cell type to the next, depending on which genes are expressed in it.

[5] Strictly speaking, transduction is a process that converts an information pattern in one modality into a comparable pattern in another, as when a telephone converts the air pressure patterns produced by the speaker into electrical signals, and then converts them back to air pressure patterns for the listener. This does not happen in signal transduction, since there is no direct correspondence between the information cells detect and the on/off responses they generate.

[6] The brain is not just composed of neurons, since it contains an equal number of *glial cells*, such as astrocytes and oligodendrocytes. Glia were initially only thought to take care of support functions in the brain, like providing structural scaffolding, myelinating axons, and combating pathogens. It now appears that they also play a part in information processing by modulating the synaptic transmission of neuronal signals in some parts of the brain.

References

Black IB (1991) Information and the Brain: A Molecular Perspective. MIT Press, Cambridge

Cairns-Smith AG (1996) Evolving the Mind: On the Nature of Matter and the Origin of Consciousness. Cambridge University Press, Cambridge

Cornish-Bowden A (2004) The Pursuit of Perfection: Aspects of Biochemical Evolution. Oxford University Press, Oxford

Dulbecco R (1987) The Design of Life. Yale University Press, New Haven

Glynn I (1999) An Anatomy of Thought: The Origin and Machinery of the Mind. Oxford University Press, New York

Harold FM (2001) The Way of the Cell: Molecules, Organisms and the Order of Life. Oxford University Press, New York

Holcombe M, Paton R (1998) Information Processing in Cells and Tissues. Plenum Press, New York

Koch C (1999) Biophysics of Computation: Information Processing in Single Neurons. Oxford University Press, New York

Küppers B-O (1990) Information and the Origin of Life. MIT Press, Cambridge

Loewenstein WR (1999) The Touchstone of Life: Molecular Information, Cell Communication, and the Foundations of Life. Oxford University Press, New York

Monod J (1971) Chance and Necessity. An Essay on the Natural Philosophy of Modern Biology. Alfred P. Knopf, New York

Niehoff D (2005) The Language of Life: How Cells Communicate in Health and Disease. Joseph Henry Press, Washington

Noble D (2006) The Music of Life: Biology Beyond the Genome. Oxford University Press, Oxford

Stevens CF (1995) The cellular basis of learning and memory. In RD Broadwell (Ed), Neuroscience, Memory, and Language. Library of Congress, Washington

Weisskopf VF (1966) Knowledge and Wonder: the Natural World as Man Knows It. MIT Press, Cambridge

Whitfield JC (1984) Neurocommunications: An Introduction. John Wiley & Sons, Chichester

Chapter 13
Genetic Messages

Abstract Every living organism inherits the information it needs to develop from a microscopic speck into a fully formed member of its own species, and then passes this on to its offspring so they can do the same. The information the genes convey is encoded in the sequence in which the nucleotides are arranged on the DNA molecules, for this specifies the sequence in which various amino acids become linked to form particular proteins. The information contained in the genome is necessary, but not sufficient, for creating an organism, since growth and development also depend on feedback from the milieu in which they take place. The interaction between nature and nurture is really an interaction between inherited and acquired information, for development is an information-driven process, not a physically driven one.

Every living organism inherits the information it needs to develop from a microscopic speck into a fully formed member of its own species, and then passes this on to its offspring so they can do the same. The messages contained in the genes of developing organisms enable their cells and organs to end up in their proper places, their regulatory mechanisms to maintain their physical integrity, and their sensory and motor mechanisms to let them adapt to changes in their environment. Every cell and organism reproduces itself by passing on information from its own genome to its offspring's, either asexually by a single cell dividing into two, or sexually, by mixing genetic material from two different parent cells. While asexually produced cells have exactly the same genetic makeup as their parents (i.e., they are clones), sexually produced ones have their own unique array of genes because of the way the information from their parents is mixed together. Sexual reproduction is a major source of the variation on which natural selection operates, since some individuals end up being genetically better able to adapt and prosper. Evolution is essentially a process of selecting individuals whose inherited information enables them to outcompete other members of their own species in terms of reproductive success and survival (Ridley 2003).[1]

A. Reading, *Meaningful Information: The Bridge Between Biology, Brain, and Behavior*,   105
SpringerBriefs in Biology 1, DOI 10.1007/978-1-4614-0158-2_13,
© Springer Science+Business Media, LLC 2011

The debate about the relative importance of nature (inherited information) and nurture (acquired information) in determining the various traits of plants and animals has largely been superseded by a realization that the two interact continuously through every stage of the life cycle. How and when most genes have an effect depends on their interactions with other genes and the contexts in which they operate. The vast majority of the traits and abilities that individuals possess (their *phenotype*) cannot be predicted from the individual genes they possess (their *genotype*), since there is no direct one-to-one correspondence between them. With a few exceptions, like eye color and Huntington's disease, how we turn out is determined by a series of recurring interactions between the messages our genes carry, the milieu in which they get expressed, and the information we derive from the environment.[2]

Genes

The genetic material in a cell's nucleus is generally arranged as two matching strands of DNA (deoxyribonucleic acid) that are connected by a series of ladder-like struts and entwined around each other in a double helix pattern to form the chromosomes. Every multicellular species has a given number of paired chromosomes, one of which is inherited from each parent. Humans, for instance, have 23 pairs of chromosomes, while domestic cats have 19 and house mice 30. The "struts" that hold the two strands together are made up of complementary pairs of chemical nucleotides. The nucleotide adenine (A) is paired with the nucleotide thymine (T), and the nucleotide guanine (G) with the nucleotide cytosine (C), so that there are four possible ways they can link the DNA strands together: A<>T, T<>A, G<>C, and C<>G. The nucleotide sequence on one strand of DNA is always the exact compliment of the sequence on the other strand, since each nucleotide can only pair with its designated partner. Thus, although each strand has a different sequence of nucleotides, they both contain exactly the same information (Rose 2005, p. 16).[3]

The information the genes convey is encoded in the sequence in which the nucleotides are arranged along the strands of DNA, for this is what determines which protein a particular section of DNA can make (Maynard-Smith 2000). Each gene consists of a sequence of several hundred nucleotides, with each set of three consecutive nucleotides constituting a single unit of genetic information called a *codon*. The order in which the nucleotides are arranged in each codon and the order in which the codons are arranged along each gene specify the sequence in which various amino acids get linked together to form particular proteins. However, only a small percentage of our DNA is used to encode the approximately 23,000 genes that constitute our human genome. The rest of our DNA consists of noncoding sequences that, among other things, regulate and modify the activity of the genes. While these noncoding nucleotide sequences were not initially thought to contain any meaningful information, it is now clear that they play an important role in determining when and where particular genes get expressed.[4]

The information encoded in the genes determines the chemical structure of the various proteins through the two-step process of *transcription* and *translation*.

When a particular gene is activated in a cell's nucleus, the information encoded in it is first *transcribed* from one of the DNA strands to a single strand of messenger RNA (ribonucleic acid). This is then transported to the ribosomes in the cytoplasm, where it is *translated* into the amino acid sequence of a given protein. Each three-nucleotide sequence of RNA (i.e., each codon) represents one of the 20 amino acids from which proteins are made, so that the composition and order of the codons determine the composition and order of the several hundred amino acids that form the protein molecule (Harold 1986). The codon guanine–adenine–guanine (GAG) on an RNA strand, for instance, is translated by the ribosomes into the amino acid glutamine, and the codon cytosine–guanine–guanine (CGG) into the amino acid arginine.

The various genes and proteins that are expressed in a given cell at a given time direct and determine its functioning at that time. Which genes are turned on or turned off in a cell is controlled, however, by the regulatory sections of the genome and by certain proteins that function as transcription factors. Transcription factors can either permanently shut down certain genes, such as kidney cell genes in non-kidney cells, or temporarily regulate the activation of other ones, including those that control metabolism and growth (Pennisi 2007). The information stored in the genome does not actually tell the cells what to do; it only tells them what proteins to manufacture at a particular time, and these proteins then detect and transmit the signals that regulate what the cell does and when it does it.

Growth and Development

Although the puzzle about how a fertilized egg (*zygote*) can transform itself into a complex multicellular organism is clearer today than it was in Darwin's time, it is no less amazing. How does this microscopic cell know which tissues and organs to form, where to place them, and what kind of animal to turn into? Although it seems to be following some sort of internal blueprint, there really is not one; it is simply a self-organizing, information-driven process in which each step leads to the next one until a final end point is reached. The zygote divides into two cells, and they divide into four, and these into eight, and those into 16, by which time it has become a little cellular ball, with some cells on the inside and some on the outside. Each of these cells has an identical set of the organism's genes, but the cells on the inside and the ones on the outside now have different environments, and that is what seems to set them on the differing pathways they will take. The developing embryo's cells detect and respond to signals from each other and from their immediate surroundings by switching on or off specific genes at specific times, some of which are only active during development. The particular genes expressed in a cell then mold and shape the growth of the developing tissues and organs (Coen 1999).

The information contained in the genome is necessary, but not sufficient, for turning a fertilized egg into an organism. Growth and differentiation also depend on cues and signals from both the external environment in which the embryo develops and the internal milieu that it itself creates. As growth and development proceed, cells lose their ability to express certain genes and become locked into the particular patterns that define their ultimate functions and then pass these latter patterns on to their progeny. In this way, the various cell types become differentiated from each other by which genes get turned on and off inside them and, thus, by which proteins they make. The different types of cells migrate along various chemical pathways to their appropriate destinations in the organism, much like the way *E. coli* cells detect and "swim" along certain chemical gradients in their surroundings.[5]

The degree of similarity between the molecular mechanisms that regulate development in different life forms is truly remarkable. They are all fashioned from the same ingredients and according to the same general principles and even by many of the same genes. Evolutionary changes have not only been brought about by forming new genes through mutations and copying errors, but also by using old ones in new and different ways. The same set of genes, for instance, regulates eye development in a host of different organisms, including insects, mollusks, and vertebrates. What is critical in determining the distinctive features of the different organisms, however, is not just which genes are involved, but the timing of when they get turned on, in what sequence and combination, and for how long. The reason why there is so much overlap between the genes possessed by different species is that they do not use all of them in exactly the same way or for exactly the same purpose. Evolution is a tinkerer that makes new structures out of old ones by modifying just where, when, and how various genes get expressed during development.[6]

Development and evolution are essentially information-driven processes that cannot be explained by the ordinary laws of physics. The critical ingredient passed from parent to offspring is the meaningful information encoded into parts of its DNA, since this is what spells out the steps that have to be taken to transform the embryo into a fully developed member of its species. Some of the information conveyed by the nucleotide patterns in the genome is the same for every member of a given species, since it is what ensures that offspring and parents are of the same kind (i.e., that a rabbit embryo will turn into a rabbit). Some inherited information differs among members of the same species, however, so they do not all grow up to have identical traits. These variable aspects of inherited information are the ones that natural selection operates on to mold the particular species, because the ones that favor an individual's reproductive fitness become increasingly represented over time in a species' gene pool. Genetic and cultural evolution are alike in this respect, since both result from the differential rates of transmission of beneficial information from one generation to the next.[7]

Notes

[1]Gould and Lewontin (1979) point out that there are constraints on the evolutionary adaptations that can occur because inherited traits are not independent units that can change without affecting other aspects of an organism's structure or function. Nor are all evolutionary changes adaptive, since some are unavoidable by-products of genetic changes that have been selected for unrelated reasons. As Jablonka and Lamb (2005) note, the genome is like a miniature ecological system in which changes in one gene affect the functioning of others. This is why a gene that has a certain effect in

one species may have a different effect in another, especially in sexually reproducing species. Natural selection *selects* the entire genomes of "fit" individuals, not just the particular genes that give them a reproductive advantage.

[2]Trying to assess the relative contributions of nature and nurture to the development of particular traits in individuals is generally futile, since neither can have an effect without the other. It is possible, however, to make a statistical estimate of the likely contribution of certain genes to certain physical characteristics *in a given population*. The National Cancer Institute, for instance, estimates that U.S. women who have inherited abnormal BRCA1 or BRCA2 genes have about a 60% lifetime chance of developing breast cancer, compared to a 12% risk in women who do not have these mutations (http://www.cancer.gov). The normal BRCA1 and BRCA2 genes help suppress tumor growth by repairing damaged DNA sequences, something the abnormal forms are apparently unable to do.

[3]A more detailed account of molecular genetics is available at Gardel et al. (2004).

[4]There does not appear to be any clear relationship between the number of chromosomes or genes a species has and its degree of complexity. The number of different proteins an organism can make can be greater than the number of genes it has, since some genes carry the code for making several different proteins. Hobert (2008) believes that the complexity of a particular organism may be related to the complexity of the ways in which the nonprotein coding parts of its DNA regulate the expression of its designated genes.

[5]Evolutionary developmental biology (evo-devo) is a discipline formed by linking developmental and evolutionary biology. Evo-devo compares the developmental processes of different animals and plants, studies how these processes evolved, and attempts to determine the ancestral relationships between them. It explores the molecular and genetic processes that regulate embryonic development and how their modification can lead to the emergence of novel features and new species (Carroll 2005).

[6]In an illustrative study, Halder et al. (1995) transferred the *eyeless* gene that is normally involved in eye development in mice into a fruit fly (*Drosophila*), and then activated it in one of its legs. Amazingly, although an eye developed at the new site, it was a fruit fly eye, not a mouse one. This suggests that the same gene conveys information about where to place an eye in both species, but not what eye to place there.

[7]The parts of the genome that are the same for every member of a species are responsible for the between-species aspects of evolution since they contribute to the relative fitness of species that compete with or prey on one another.

References

Carroll SB (2005) Endless Forms Most Beautiful: The New Science of Evo Devo and the Making of the Animal Kingdom. WW Norton, New York

Coen E (1999) The Art of Genes: How Organisms Make Themselves. Oxford University Press, New York

Gardel C, Lander E, Weinberg R, Chess A (2004) Introduction to Biology. MIT OpenCourseWare, http://ocw.mit.edu

Gould SJ, Lewontin RC (1979) The Spandrels of San Marco and the Panglossian Paradigm: A Critique of the Adaptationist Programme. Proc. Royal Soc. London, Series B, 205 (1161): 581–598

Halder G, Callaerts P, Gehring WJ (1995) Induction of ectopic eyes by target expression of the *eyeless* gene in *Drosophila*. Science 267: 1758–1791

Harold FM (1986) The Vital Force: A Study of Bioenergetics. W.H. Freeman, New York

Hobert O (2008) Gene regulation by transcription factors and microRNAs. Science 319: 1785–1786

Maynard-Smith J (2000) The concept of information in biology. Philosophy of Science 67: 177–194

Pennisi E (2007) A new window on how genes work. Science 316: 1120–1121

Ridley M (2003) Nature Via Nurture: Genes, Experience, and What Makes Us Human. HarperCollins, New York

Rose S (2005) The Future of the Brain: The Promise and Perils of Tomorrow's Neuroscience. Oxford University Press, Oxford

Chapter 14
Feelings as Information

Abstract Feelings convey meaningful information about the state of our internal milieu and our relationship with the external world. Feelings are subjective qualities we attribute to the objects and events we perceive, since there is nothing inherently good or bad (or right or wrong, ugly or beautiful, etc.) about the patterns of energy and matter we detect. The various types of feelings we experience include qualia, somatic sensations, mental state appraisals, moral sentiments, esthetic feelings, and emotions. Emotions convey qualitative types of evaluative information about the world around us, which helps us navigate our way through it. All living creatures have a built-in imperative to behave in ways that maximize pleasant feeling states and minimize unpleasant ones in order to promote their own and their species' survival.

Our feelings convey meaningful information about the state of our internal environment and our relationship with the external one. They are subjective sensations that we *feel*, just like odors are ones we smell and sounds are ones we hear. Feelings differ from the other sensations we experience because they represent *evaluations* of the patterns of energy and matter that we detect. We do not just experience incoming patterns of sensory stimuli; we also compare them to internal reference patterns that we have developed, either as a result of evolution or learning. What we experience as *feelings* are simply the outputs of these analog-type comparisons. We screen incoming patterns of sensory input on several fronts, including evaluating whether they are similar to previous ones (feel strange or familiar), whether they are likely to benefit or harm us (feel good or bad), and whether they are consistent with our existing beliefs (feel right or wrong). While all sentient animals have at least some awareness of how they feel, since this is how sentience is defined, non-sentient species do not—they just act on this type of evaluative information automatically, without any awareness of what they are doing, or why they are doing it.

The various types of sensory information that organisms distinguish are essentially *value-free*, since there is nothing inherently good or bad about the patterns of energy and matter they detect. There is nothing in the entire universe, in fact,

A. Reading, *Meaningful Information: The Bridge Between Biology, Brain, and Behavior*, 111
SpringerBriefs in Biology 1, DOI 10.1007/978-1-4614-0158-2_14,
© Springer Science+Business Media, LLC 2011

that seems to be inherently good or bad, bright or dull, ugly or beautiful, or the like, other than in the "eyes" of some sort of beholder. Feelings are subjective qualities we attribute to the objects and events we perceive, not properties of the objects or events themselves. Animals evaluate the sensory input they receive, either consciously by how it makes them feel, or nonconsciously through processes that trigger automatic responses. As Johnston (1999) notes: "the environment is filled with lawful and consistent events, but such happenings are devoid of all meaning, for the laws of physics and chemistry that govern the behavior of these events have no inherent purpose or intention." The evaluative process associated with feeling states is a fundamental part of the circuitry that enables animals to respond and adapt to what they encounter.

Feelings (and the equivalent in non-sentient species) typically provide information about *values*, i.e., appraisals about whether a detected object or event is likely to benefit or harm the organism. Good and bad, right and wrong, and true and false are subjective values we apply to our perceptions, rather than qualities of the matter and energy patterns that elicit them. Values, in the sense used here, are initially based on built-in reference standards that have evolved because of the way they have enabled different species to detect information that increases their evolutionary fitness. Objects and events that benefit a given species' fitness become associated with positive feeling states as a result of natural selection, while ones that harm it become associated with negative ones. What is good for one species or individual, however, may not be good for another, since each has its own set of needs and values. As Dolan (2002, p. 1191) notes: "Value refers to an organism's facility to sense whether events in its environment are more or less desirable. Within this framework, emotions represent complex psychological and physiological states that, to a greater or lesser degree, index occurrences of value."

While we refer to the conscious outcomes of these analogical comparisons as *feelings*, there is no equivalent name for them in non-sentient species. Because of the obvious continuity in the way that sentient and non-sentient species evaluate sensory input, the terms *feeling state* and *value* are used here to apply to all living entities, whether or not they are experienced consciously. As Nagel (1974) points out, we do not know what other species experience, although we can make inferences about what they sense and how they feel from how they behave. We simply assume that any sensory input that causes an organism to move toward its source or increase what it is doing has a positive (good) value, and any input that causes it to move away from its source or decrease what it is doing has a negative (bad) value. This enables us to operationally determine how other animals evaluate various stimuli, even though we do not have access to the internal states they elicit.[1]

Darwin (1872) was taken by the similarities in the way humans and certain other species express their feelings. He noted, for instance, that fear is manifested in virtually all mammals by a widening of the eyes and mouth, trembling muscles, hair standing on end, chattering teeth, and loosened sphincters. These responses are why we generally assume that mammals have an awareness of feelings like pleasure and pain, and that most invertebrates do not. Species that do not have a central nervous system are probably unable to be aware of their internal states or generate responses that do not involve automated reflexes. Species that do have a central nervous system are able to supplement their inborn evaluative abilities with ones they learn from experience, so that they end up with a variety of subjective standards for evaluating how the objects and events they encounter make them feel.

The Types of Feelings

The various kinds of feelings we experience represent the output of appraisal processes that provide us with meaningful information about the "value" of the external objects and events we detect, as well as about our internal functioning. These appraisals involve a type of goodness-of-fit comparison between what is being experienced and some type of internal yardstick. The informational output of these comparisons is analogic in nature, which is what gives feelings their ineffable, difficult-to-describe quality. They only tell us how well something matches a given standard, not how or why it does. Analogic outputs are similar to the type of information we get when we eat an ice cream or put our hand under a running faucet; we can tell whether they feel hot or cold, but not what their temperature is. While our body temperature serves as the reference standard in these two examples, we have no idea about the nature of the internal standards involved in the generation of our feelings. Some of them are obviously built in genetically, like the feeling of pleasure associated with food and sex, the feeling of fear associated with threats of harm, and the feeling of revulsion associated with bad tastes or odors. Other reference standards are learned from experience, as well as from family and societal values, like the feeling of guilt we experience when we transgress moral boundaries or the shame we sense when we fail to live up to our internalized ideals.

Feelings convey a primitive, pre-verbal type of information. Unlike the quantitative type of digital information that words and numbers provide, the informational content of feeling states is *qualitative* in nature and thus cannot be combined and re-arranged to produce rules, make generalizations, or construct models. Analog information is limited to matching one pattern against another to see how they compare—like or unlike, more or less, better or worse, etc. It is always *relative* to some sort of real or implied standard, and is thus not amenable to the kind of cognitive processes that underlie human thought and reasoning. Different types of feeling states provide different types of information, with each species having evolved ways of evaluating the type of sensory input that is of greatest benefit to it. Humans, for instance, have access to a wide variety of feeling states, each of which provides information about a different facet of our lives, as outlined below. Although the terms *feelings* and *emotions* are often used interchangeably, emotions are represented in the following as only one of the feeling states we experience.

Qualia are emergent qualities we attribute to certain perceptions, like the colors we associate with certain wavelengths of light, the sweetness we attribute to the taste of honey, and the sickening quality we ascribe to the smell of rotten eggs. They are subjective experiences that are not explainable by the physical properties of the stimuli that generate them. The beauty of a sunset, for instance, is not a property of the sunset, nor is the unpleasant taste of castor oil a property of the castor oil; they are simply subjective qualities we attribute to these physical sensations, based on our personal and evolutionary history.

Physical assessments are analogic evaluations of the sensory input generated by the physical properties of certain stimuli, rather than by information. They include sensations like how heavy or smooth or hot an object feels, how hard the wind is blowing, or how loud the orchestra is playing.

Somatic sensations are our primary source of information about how our bodies are functioning. They include negative feelings, like pain and discomfort, that indicate that something is amiss, although they do not necessarily tell us what is wrong. We also experience unpleasant sensations that inform us about physiological needs that should be addressed, including hunger, thirst, fatigue, and sleepiness. The range of positive somatic feelings is narrower, possibly because being able to detect potential trouble is more important from an evolutionary standpoint. Positive somatic sensations include the feelings of vigor and well-being we experience when we are physically fit and healthy.

Mental state appraisals provide us with information about the current state of our mental functioning. They include feelings of curiosity, interest, determination, boredom, and confusion, as well as ones of self-confidence, certitude, pessimism, and doubt. These states both shape the content of our current thoughts and are shaped by them.

Moral sentiments provide information about conformity with social and religious standards of right and wrong, and include feelings of pride, guilt, righteousness, embarrassment, humiliation, and shame. Since there are few innate determinants of what is right or wrong, moral sentiments usually involve the mental appraisal of given acts in relation to a learned system of values. These standards generally involve socially and religiously derived covenants, like the Ten Commandments and the Golden Rule. The fundamental normative and ethical values of a culture are based on such traditional belief systems, rather than scientific facts (Tiselius and Nilsson (1970). Other animals do not appear to experience a moral or ethical sense of what is right or wrong.[2]

Esthetic feelings provide information about perceived beauty and harmony in nature, art, music, and religious experience, and include feelings of serenity, wonder, awe, and reverence. It is not clear whether there are any innate models for what is esthetically pleasing, since standards of beauty vary considerably from age to age and culture to culture. Just what it is that some people find so pleasing about certain experiences can be unfathomable to others.[3]

Emotions are probably our most frequently experienced feelings, for they color how we perceive and respond to the information we detect in our day-to-day endeavors. They are our major source of information about our *relationship* with the world in which we live—about how well or poorly we are functioning in our everyday lives. As Johnston (1999) points out, emotional information is part of a regulatory system that helps us steer a safe passage through the various ups and downs we encounter. Pleasant emotions (*feeling good*) indicate that all is well with the world and energize us to maintain or increase what we are doing, while unpleasant ones (*feeling bad*) signal that something is amiss and mobilize us to change our behavior. As Bentham (1789) notes: "nature has placed mankind under the governance of two sovereign masters, *pain* and *pleasure*. It is for them alone to point out what we ought to do, as well as to determine what we shall do."[4]

The Varieties of Emotional Experience

A large number of human emotions have been identified, some associated with distinctive physiological responses, some with characteristic patterns of facial and bodily expressions, and some with various degrees of cognitive appraisal. These features have not proved to be a useful basis for defining them, however, since they do not all share them. Emotions are feelings that convey evaluative information about our interactions with the world around us, with each of them providing information about whether a particular aspect of it is likely to help or harm us. There is, however, no agreed-upon way of identifying which feeling states represent emotions or specifying the functions they serve. The following groupings illustrate the various types of information emotions can convey:[5]

Current-state emotions convey information about the current state of our relationship with the world around us, about whether or not things are currently going well or poorly for us. They include happiness, unhappiness, pleasure, sadness, dissatisfaction, contentment, enjoyment, and delight.

Affiliative emotions convey information about the desirability (or undesirability) of another person or object. They include love, hate, like, dislike, disapproval, jealousy, envy, contempt, lust, loathing, and resentment.

Defensive emotions convey information about potential harm. They include concern, worry, anxiety, apprehension, fear, alarm, and terror.

Infringement emotions convey information that our body, territory, property, beliefs, or expectations are being violated, or are in danger thereof. They include irritation, annoyance, frustration, anger, and rage.

Prospective emotions convey current information about how we see our future relationship with the world. They include excitement, elation, hope, despair, depression, and dread.

Retrospective emotions convey information about our current evaluation of aspects of our past behavior. They include guilt, regret, and remorse.

Emotions can be used to evaluate both our direct sensory experience and the scenarios we create in our mind. Being able to evaluate how imagined scenes make us feel allows us to sense the outcome of differing strategies without actually having to undertake any of them. Emotions evoked by imagery are, however, less vivid than those produced by external stimuli, possibly because they do not have to activate behavioral responses (Rolls 2005, p. 26). The feeling states experienced by the other sentient species are limited to conveying information about their current state of affairs, since they are unable to conjure up scenes about the past or imagine ones about the future.

The current-state feelings of happiness, sadness, anger, fear, disgust, and surprise are often referred to as *primary emotions* because they are expressed by people in every corner of the globe, as well as by infants and individuals with dementia. They are all accompanied by facial and bodily expressions that can have an effect on others as nonverbal cues and signals (Whybrow 1997). This group of emotions appears to be evolutionarily ancient, since it includes feeling states that other sentient species can also experience. Most of our other emotions, such as ones like envy, pride, regret, and loneliness, do not have a distinctive form of nonverbal expression and do not appear to be shared with other species. While our primary emotions are mainly processed sub-cortically (in the limbic system), most of the other ones we experience require some degree of conscious cortical appraisal.[6]

Although the information our emotions convey plays a major role in shaping how we respond to the objects and events we encounter, the system is not foolproof since our reference standards cannot always be relied on to assess longer-term benefit or harm. Some things that feel good in the present, like illicit drugs and eating more than we need, may not be good in the long run, while some that feel bad at the time, like exercise and studying, may be of benefit later on. Choosing between immediate and deferred sources of gratification is an age-old dilemma that involves deciding between on-line and off-line value assessments. Although pursuing immediate pleasure and avoiding immediate pain are innately attractive, especially as they can be pursued without too much thinking, successful adaptation also involves investing in the future in ways that increase the overall amount of positive feelings during a person's lifetime.

Behavioral Regulation

All living creatures have a built-in imperative to behave in ways that maximize positive feeling states and minimize negative ones in order to promote their own and their species' well-being. This has been brought about by natural selection, since individuals who are best able to associate agreeable feelings with things that benefit them, and disagreeable feelings with ones that harm them, are more likely to prosper and pass these traits on to their offspring. Every species differs, however, in what makes it feel good and what makes it feel bad (Damasio 1999). The basic strategy is illustrated in the following diagram: pleasurable feelings provide feedback that maintains or increases current behavior, while unpleasant ones provide feedback that decreases or changes it (Fig. 14.1).

Similar mechanisms regulate the behavior of non-sentient species by automatically activating or deactivating preprogrammed responses when certain input patterns are detected, in much the same way that a thermostat can automatically regulate the temperature without having any awareness of what it is doing. When *E. coli* senses a toxic substance, for instance, it tumbles away randomly until it ends up in an environment where it can no longer sense it. Sentient animals, on the other hand, generally have some degree of choice about how to respond to their feelings,

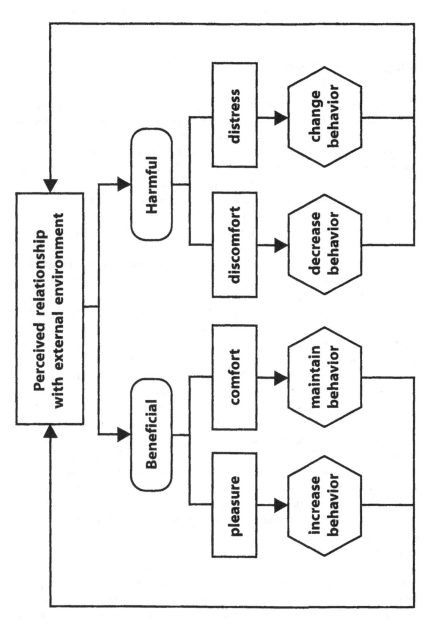

Fig. 14.1 The regulation of behavior by emotional information (Reprinted from Reading (2004) with permission of The Johns Hopkins University Press)

since most sentient states are only associated with a *disposition* to respond in a certain way, rather than a specific way of doing it (Frijda 2006). We usually respond to the unpleasant feeling of hunger, for instance, by choosing where, when, and what to eat, and to the feeling of satiation by when to discontinue our intake.

Thinking Versus Feeling

Thoughts and feelings are often viewed as being opposed to each other, one logical and rational and the other illogical and impulsive. Thinking is mainly based on the manipulation of language-based symbols according to certain rules and conventions, while feelings are outcomes of goodness-of-fit comparisons. Decision making in the former involves logically weighing the pros and cons of various options, while in the latter imagined scenes are compared to see which one feels best. These are, however, just complementary ways of processing information, one managed primarily by digital operations that deal with facts and relationships, and the other by analog ones that assess values and qualities. Feeling states (and their nonconscious equivalents) are the more primitive, having some form of representation across the entire spectrum of animal life, while thinking is essentially limited to vertebrate species. As Hauser (2006, p. 31) notes: "Intuitions and conscious reasoning have different design specs. Intuitions are fast, automatic, involuntary, require little attention, appear early in development, are delivered in the absence of principled reasons, and often appear immune to counter-reasoning. Principled reasoning is slow, deliberate, thoughtful, requires considerable attention, appears late in development, justifiable, and open to carefully defended and principled counterclaims."[7]

Our conscious world is made up of both incoming sensory information and self-generated thoughts and images. We evaluate these converging data streams during our waking hours in order to ascertain their meaning, assess how they make us feel, and determine whether they are consistent with what we believe. The vast majority of these evaluations proceed automatically, completely outside of our awareness. They guide us effortlessly through the myriad of everyday choices we face, like what to wear, eat, or buy, where to do it, how to get there, and what to say when we do. These decisions, once they become routine, are based on nonconscious assessments that automatically link the incoming information to its designated response. We are aware of the choices we make, but not of how we make them; we just know they *feel* the right thing to do.

Novel situations, unfamiliar objects, and sensed discrepancies quickly receive our conscious attention and alert our decision-making processes. We can choose to base our response either on our emotional reaction to a particular situation or our rational analysis of it. Although the Enlightenment enthroned reason as our highest virtue, our instinctive "gut decisions" can sometimes be more accurate than rational deliberation (Gladwell 2005). It seems that conscious cost/benefit assessments are generally better with decisions that deal with quantitative issues or involve a limited number of variables (like which college course to take), while intuitive, feelings-based ones are often better with ones that deal with qualitative assessments or involve complex variables (like which car to buy). Dijksterhuis et al. (2006) maintain that, because conscious thought is rule-based, it is unable to handle value judgments that involve a large number of variables as effectively as the pattern-based evaluations that underlie intuitive choices. Both logical and intuitive decisions depend, however, on the accuracy of the relevant knowledge and understanding individuals possess, no matter whether the assessment is made consciously or nonconsciously. That is why the "gut decisions" of well-informed individuals are more likely to be correct than those of less knowledgeable ones.[8]

Our ability to evaluate how imagined scenarios make us feel greatly expands the amount of meaningful information available to us, since it allows us to determine which options are likely to benefit or harm us, both in the short and the long term. As Gilbert and Wilson (2007, p. 1352) point out: "simulations allow people to 'preview' events and to 'prefeel' the pleasures and pains those events will produce." We can, for instance, imagine what we would feel like if went on a vacation, got married, or landed in jail, and shape our behavior accordingly. We can also sense whether we would later regret taking a certain course of action (or would regret not taking it), and use this information in our decision making. We can choose what to do either through logical thought, like deciding not to smoke because of the risk of lung cancer, or by picturing various options in our mind's eye and assessing how each would make us feel, like imagining a grave full of cigarette packs. Feelings provide value assessments that are not available in logical thought, while thinking has access to facts and rules that are hidden from analogic appraisals. This is why when we read a novel we picture the story in our mind in order to gain access to the emotions involved. As Calne (1999) points out, reason "has improved *how* we do things, but has not changed *why* we do things," and Simon (1983, p. 7) adds: "reason can't tell us where to go: at best, it can tell us how to get there."

Many people fail to see their feelings as valuable sources of information; they just experience them without trying to discern the information they contain. By tuning in to what they are telling us, however, we can get an idea of how our evaluative systems are functioning and identify aspects that may need to be modified. Negative feelings, for instance, rather than just being states to avoid or suffer though, often offer opportunities for learning about our underlying beliefs and assumptions. Whether we are controlled by our emotions or control them depends largely on whether we understand the information they convey. So-called *emotional intelligence* involves the ability to monitor our own and others' emotions, discriminate among them, and use this information to guide our thoughts and actions (Ciarrochi et al. 2001). People differ a great deal, however, in how effectively they can identify, express, understand, and regulate their feelings. Impulsive individuals are generally unable to reflect on the information their emotions contain, while compulsive ones tend to discount feelings and let their thoughts control what they do. Both extremes impair adaptive functioning by limiting the amount of meaningful information available to the individuals involved.

Notes

[1] The term *value* is used in everyday language to refer to many other types of evaluations, not just ones that generate feeling states. We have family values, economic values, corporate values, and so on, most of which are associated with cognitive processes rather than sensory perceptions, although they all involve some type of analogic matching. In order to distinguish it from the others, the type of evaluative information that we detect and respond to is sometimes referred to as *valence*, rather than *value*, and the associated subjective states as *hedonic tones* rather than *feelings*.

[2] The taboos against incest and physically harming members of one's own social group seem to have a genetic basis in humans.

[3] Hauser (2006, p. 31) quotes Hume (1739): "Beauty is not a quality of the object, but a certain feeling of the spectator, so virtue and vice are not qualities in the person to whom language ascribes them, but feelings of the spectator."

[4] There is little agreement about which feeling states ought to be considered emotions. Denton (2006), for instance, includes bodily states like pain, thirst, and nausea as emotions. Solomon (2007) does not believe that emotions should be thought of as feelings, since he sees feelings as originating solely from bodily states, like the way blushing occurs. Damasio (2003) differentiates feelings from emotions, with the former being the subjective experience and the latter the entire process.

[5] There are a number of theories about what emotions are and what they do, but none conceptualizes them primarily as sources of information or explicitly recognizes their continuity with the other feeling states. The approach taken here is, however, consistent with Ortony et al. (1988), who propose that "the primary function of emotion is to provide information about some special harm or benefit that inheres in the relationship between the person and the environment;" with Damasio (1994), who notes that "feelings are the sensors for the match or lack thereof between our nature (inherited and acquired) and our circumstances;" and with Nussbaum (2001), who states that emotions "are judgments of value which appraise whether external objects are salient for our own well-being." The concept of emotions as information is also discussed in Reading (2004). A number of other theories have been proposed, including ones by Solomon (1976), Frijda (1986), De Sousa (1987), McNaughton (1989), Oatley (1992), LeDoux (1996), and Plutchik (2001). Overviews of these differing approaches can be found in Eckman & Davidson (1994), Lewis & Haviland-Jones (2000), Evans (2001), and Fox (2008).

[6] Incoming sensory information is relayed from the thalamus to the limbic system, where it can activate the hypothalamus and the autonomic nervous system in ways that enable "emergency" emotions like fear and anger to generate rapid responses. It is also transmitted from the thalamus to the sensory cortex, and then to the association and prefrontal regions for conscious appraisal. These latter areas initiate behavioral responses through their connections to the frontal motor cortex (LeDoux 1996).

[7] Some individuals apparently think more in pictured symbols than in verbal ones. Einstein (1954, p. 25), for instance, maintained that: "The words of the language, as they are written or spoken, do not seem to play any role in my mechanism of thought. The physical entities that seem to serve as elements in thought are certain signs and more or less clear images which can be 'voluntarily' reproduced and combined." Grandin (1995, p. 29), an autistic scientist, believes that most autistic individuals have problems learning things that cannot be thought about in pictures.

[8] Several authors have recently described how intuitive decisions can be better than ones based on conscious reasoning, including Gladwell (2005), Gigerenzer (2007), Ariely (2008), and Lehrer (2009). The claims that this dethrones reason from its pedestal of superiority are not entirely valid, however, since conscious reasoning is the main way we establish the knowledge base and internal models that serve as internal references for making intuitive decisions.

References

Ariely D (2008) Predictably Irrational: The Hidden Forces That Shape Our Decisions. HarperCollins, New York

Bentham J (1789, 1970) An Introduction to the Principles of Morals and Legislation. J H Burns, HLA Hart (Eds.). The Athlone Press, London

Calne D (1999) Within Reason: Rationality and Human Behavior. Pantheon Books, New York

Ciarrochi J, Forgas JP, Mayer JD (2001) Emotional Intelligence in Everyday Life: A Scientific Inquiry. Psychology Press, Philadelphia

Damasio AR (1994) Descartes' Error: Emotion, Reason, and the Human Brain. GP Putnam's Sons, New York

Damasio AR (1999) The Feeling of What Happens: Body and Emotion in the Making of Consciousness. Harcourt Brace, New York

Damasio AR (2003) Looking for Spinoza: Joy, Sorrow, and the Feeling Brain. Harcourt, Orlando

Darwin C (1872,1934) The Expression of the Emotions in Man and Animals, Abridged edn. Watts, London

Denton D (2006) The Primordial Emotions: The Dawning of Consciousness. Oxford University Press, New York

DeSousa R (1987) The Rationality of Emotion. MIT Press, Cambridge

Dijksterhuis A, Bos MW, Nordgren LF, van Baaren RB (2006) On making the right choice: The deliberation-without–attention effect. Science 311: 1005–1007

Dolan RJ (2002) Emotion, cognition, and behavior. Science 298: 1191–1194

Eckman P, Davidson RJ (1994) The Nature of Emotion. Oxford University Press, New York

Einstein A (1954) Ideas and Opinions. Crown, New York

Evans D (2001) Emotion: The Science of Sentiment. Oxford University Press, New York

Fox E (2008) Emotion Science: Cognitive and Neuroscientific Approaches to Understanding Human Emotions. Palgrave Macmillan, Houndmills

Frijda NH (1986) The Emotions. Cambridge University Press, Cambridge

Frijda NH (2006) The Laws of Emotion. Lawrence Erlbaum, Hillsdale

Gigerenzer G (2007) Gut Feelings: The Intelligence of the Unconscious. Viking, New York

Gilbert DT, Wilson TD (2007) Prospection: Experiencing the future. Science 317: 1352–1354

Gladwell M (2005) Blink: the Power of Thinking Without Thinking. Little Brown, New York

Grandin T (1995) Thinking in Pictures: And Other Reports from My Life with Autism. Doubleday, New York

Hauser M (2006) Moral Minds: How Nature Designed Our Universal Sense of Right and Wrong. HarperCollins, New York

Johnston VS (1999) Why We Feel: the Science of Human Emotions. Basic Books, New York

LeDoux J (1996) The Emotional Brain. Simon & Schuster, New York

Lehrer J (2009) How We Decide. Houghton Mifflin Harcourt, New York

Lewis M, Haviland-Jones JM (2000) Handbook of Emotions, 2nd edn. Guilford Press, New York

McNaughton N (1989) Biology and Emotion. Cambridge University Press, Cambridge

Nagel T (1974) What is it like to be a bat? Philosophical Review LXXXIII: 435–450

Nussbaum MC (2001) Upheavals of Thought: The Intelligence of Emotions. Cambridge University Press, Cambridge

Oatley K (1992) Best Laid Schemes: The Psychology of Emotions. Cambridge University Press, Cambridge

Ortony A, Clore GL, Collins A (1988) The Cognitive Structure of Emotions. Cambridge University Press, New York

Plutchik R (2001) The nature of emotions. American Scientist 89: 344–350

Reading AJ (2004) Hope and Despair: How Perceptions of the Future Shape Human Behavior. Johns Hopkins University Press, Baltimore

Rolls ET (2005) Emotions Explained. Oxford University Press, New York

Simon HA (1983) Reason in Human Affairs. Stanford University Press, Stanford

Solomon RC (1976) The Passions. Notre Dame University Press, Notre Dame

Solomon RC (2007) True to Our Feelings: What Our Emotions Are Really Telling Us. Oxford University Press, New York

Tiselius A, Nilsson S (1970) The Place of Value in a World of Facts (Nobel Symposium). Wiley, New York

Whybrow PC (1997) A Mood Apart: Depression, Mania, and Other Afflictions of the Self. Basic Books, New York

Chapter 15
Consciousness

Abstract Consciousness is comprised of two distinct experiences: sentient consciousness, an awareness of one's current perceptions and feelings, and reflective consciousness, an awareness of one's own self as a unique individual who can initiate actions, ideas, and images that are independent of ongoing sensory input. Most of the information processing in the natural world occurs nonconsciously, both in species that are limited to this level and in ones that are not. Consciousness is primarily about choice. While non-sentient species respond automatically to detected information, sentient ones have a degree of choice about how they respond, based on other current inputs. Reflective consciousness offers our species a much greater range of choices because of our ability to understand the meaning of detected information and envision the consequences of our actions.

Consciousness involves two distinct experiences. These are referred to here as (a) *sentient consciousness*, an awareness of one's current perceptions and feelings, and (b) *reflective consciousness,* an awareness of one's own self as a unique individual who can initiate actions, ideas, and images that are independent of ongoing sensory input. Sentient consciousness, the less highly developed of the two, is limited to experiencing analog-type information generated by sensory input and feeling states, while reflective consciousness involves an awareness of digital-type information, like thoughts, images, and language. Because it involves symbolic information, reflective consciousness is essentially limited to humans, while sentient consciousness is present in other mammals and possibly birds. It is not clear whether fish or reptiles possess any form of sentient awareness, although plants and invertebrate animals certainly do not seem to. Whether other animals experience consciousness depends on how it is defined, since the term is often used to refer to different forms of awareness. The above distinction between sentient and reflective consciousness is similar to the ones made by Edelman (1992) between *primary* and *higher order consciousness,* Mandler (1997) between *consciousness* and *reflective consciousness*, and Damasio (1999) between *consciousness* and *extended consciousness.*[1]

A. Reading, *Meaningful Information: The Bridge Between Biology, Brain, and Behavior*,
SpringerBriefs in Biology 1, DOI 10.1007/978-1-4614-0158-2_15,
© Springer Science+Business Media, LLC 2011

What It Is For

Biology searches both for proximate causes that explain *how* entities operate, as in molecular biology, and ultimate ones that explain *why* they function the way they do, as in evolutionary biology (Mayr 1961). Although no one understands how the human brain generates conscious thoughts and feelings, there are a number of theories about why it does it. Freeman and Herron (2007), for instance, believe that the various aspects of consciousness have been produced by natural selection because they afforded an evolutionary advantage to their possessors. Greenfield (1995) contends that consciousness acts as a kind of gatekeeper that makes sure that only one of the various sources of information we are processing at a given time gets our full attention, since consciousness is a serial process (while most of the brain's other operations occur in parallel). Tononi and Edelman (1988) maintain that one of consciousness's main functions is to integrate neural activities that take place in different parts of the brain, so that a response can be generated on behalf of the animal as a whole. Input that affects only a part of an animal, such as reflexes, muscular coordination, or blood pressure regulation, is generally dealt with on an automatic, nonconscious basis, just as it is in non-sentient species.

The adaptive function of both sentient and reflective consciousness appears to be to enable animals to vary how they respond to the information their sensory receptors detect, rather than just react automatically. While sentience allows individuals to consider certain contextual factors in determining how to respond to a given object or event, reflective awareness also lets them interpret the significance of the information being processed in light of their past experience and future goals. Consciousness is like a module for variation inserted between an animal's detection and response systems, with the more elaborate forms of conscious awareness leading to greater and more effective behavioral variability. Although we cannot tell what another animal's conscious experience is like, we can get some idea about the role it plays in its life by observing the circumstances which elicit the visible manifestations of feelings like pain and pleasure, as well as the extent to which it varies the way it responds to given stimuli in different contexts.

Sentient States

The basic elements of sentient consciousness consist of an awareness of bodily sensations, like hunger and pain, as well as an appreciation of externally generated feelings, like pleasure and fear. While the neuronal circuits that link the detection of information to the response it generates are relatively fixed in non-sentient creatures, they are open to input from other sources in sentient ones. Sentient awareness allows animals to have a degree of choice in how they respond to incoming information so that their behavioral patterns are not entirely automatic. They can respond to sensations of thirst or hunger, for instance, with types of search behavior that are shaped by their current circumstances, rather than being entirely predetermined. Sentient animals are also able to respond to painful stimuli by either escaping or attacking, depending on the situation, and can vary the manner in which they escape or attack.

Although animals that only possess a sentient type of consciousness are aware of feeling states and sensations, they cannot recall the past or generate scenarios about the future because they are unable to process symbolic information. Infants, severely demented individuals, and people consuming certain mind-altering substances are also limited to sentient forms of consciousness for much the same reason. Alcohol and marijuana, for instance, bring about a focus on the here and now that induces a disconnection from the world of worries and expectations, together with a drift away from abstract and logical reasoning (Greenfield 2000). The altered states of consciousness that mind-altering substances can induce usually involve a retreat into a sentient state in which there is a heightened experience of external stimuli, a loss of the sense of self, a diminished awareness of time, and an elimination of future-directed behavior. Most of our recreational activities represent acceptable ways of inducing a sentient state by focusing entirely on the present, such as in sailing or skiing, listening to a concert, watching a sporting event, or getting engrossed in a novel. We are conscious of our feelings and perceptions during these experiences, but of little else as we lose ourselves in what we are doing and do not notice the passage of time. Because our worries mostly involve concerns about the past and apprehensions about the future, present-focused states are generally experienced as pleasurable and relaxing.[2]

A number of traditional religious and cultural practices also involve loosening the mind's attachment to reflective levels of consciousness through practices that involve intense concentration on the present, including fasting, chanting, dancing, spinning, gazing at a mandala, or endlessly repeating a given phrase. In Zen, for instance, the intent is to stop conceptualizing while being fully awake, so that ordinary thoughts and attachments to worldly desires are placed in abeyance (Crook 1980; Austin 2006). Enhanced sentience is also involved in the state of *flow* that Csikszentmihalyi (1990 and 1997) popularized as an optimum experience that allows individuals to achieve a richer, more fulfilling, and more joyous sense of being. He described *flow* as "a state in which individuals become so immersed in the present moment that there is no room in their awareness for conflicts, distracting thoughts, or irrelevant feelings as their sense of self-consciousness disappears and hours seem to pass in minutes." Some people experience the loss of self that occurs during certain sentient states as a mystical experience characterized by feelings of indescribable serenity and a sense of being in absolute harmony with the universe. While such transcendent experiences are often portrayed as being due to achieving a higher level of consciousness, they are actually caused by a retreat to a more primitive one.

Although sentient animals have some awareness of their internal states and their surroundings, they are unaware of why they respond the way they do. Their behavior is generated by innate drives and sensory information, rather than by personal goals or understanding. The subjective state we experience when we are focused intensely on the present is probably the closest we can get to appreciating how sentient animals (and young children) experience the world. Although we are aware of our sensory input and internal feelings during these times, we act without any conscious thought about what we are doing or why we are doing it. Focusing on the present heightens our experience of our sensory input and feelings, which is why we try to shut out distracting thoughts when we try to hit a golf ball, read a book, or watch a movie. People in large crowds can also become so focused on the present that they enter into a sentient state in which they lose their sense of self as they get swept along by others (Le Bon 1903). We are unable to subsequently recall sentient experiences like hunger, pain, or happiness, although we can recall we had them, and can simulate them by imagining the objects or events that elicited them. Objects and events have to be experienced in reflective consciousness for us to be able to recall them later on.

Reflective States

Reflective consciousness has a number of distinctive components. These include: (a) an awareness of one's self as an initiator of thought and action, (b) an awareness that others have a similar sense of self, (c) an ability to recall the past, envision the future, and indulge in fantasy and imagination, (d) an ability to set future goals and make personal decisions, (e) an ability to communicate through the use of language, (f) a sense of the passage of time, (g) a sense of free will, and (h) an ability to reflect on our own thoughts and behavior. All of these require cortical structures that are able to process symbolic information, facilitate abstract thought, and create mental representations that can function as internal sources of meaningful information. Although rudimentary aspects of these abilities seem to be present in other primates and in some marine mammals, they are only fully developed in our own species, where they account for a great deal of our unique adaptive skills (Terrace and Metcalfe 2005). As Greenspan and Shanker (2004, p. 17) note: "Only human beings, as far as we know, can engage in reflective thinking. They can think about and judge their own thoughts and feelings. Many members of the animal kingdom can solve problems, but only human beings can use symbols to think reflectively."[3]

Information processing during states of reflective consciousness covers a range of cognitive activities, including thinking, analyzing, deciding, understanding, recalling, and imagining, all of which involve a sense of *self*, a feeling that there is someone directing them. But this is simply an illusion. There is no little homunculus in the brain controlling what we do, no "ghost in the machine," just a miraculous bunch of neurons acting as if there were one. Although we are conscious of our feelings and sensory input during sentient states, we only understand their meaning and feel we can control our own behavior during reflective ones. The sense that we are the authors of our own behavior, masters of our own destiny, emerges during our first 3–4 years of life as we become increasingly able to reflect on what we think and do (Damasio 2010).[4]

Reflective consciousness is what makes the behavioral responses of humans so varied and unpredictable, since they are determined by the meaning each of us assigns to detected information, not by the information itself (Monod 1971, p. 43).

Because we have different backgrounds, different values, and different goals, no two of us necessarily interpret or respond the same way to the same information. Factors such as religious beliefs, educational level, economic status, and personal history shape how we construe and respond to meaningful information, which is why different individuals can end up with different views on issues like global warming, abortion, and evolution. What we already believe, no matter whether right or wrong, shapes what we perceive, what we expect, and what we recall—in ways that we ourselves are usually unaware. This is why witnesses can have different recollections of the same event, why inert placebos can affect physical symptoms, and why we tend to seek confirmation of our views, rather than enlightenment.

Despite the many advantages reflective consciousness provides, it is not an unmitigated blessing, since the thoughts and images it generates can also cause anxiety and distress. While other species are only able to be fearful of clear and present dangers, we are free to worry about any number of imagined disasters just waiting to befall us, like having incurable cancer, suffering a fatal heart attack, or becoming demented. Although it confers many gifts, our unique form of consciousness also makes us aware of our frailties and inevitable mortality, burdens that other species do not have to carry. Most growing children have little awareness of these problems, since their reflective consciousness is not yet fully developed, which is probably why they seem more lighthearted and carefree than their parents.[5]

Processing Levels

Living organisms process meaningful information in one of three ways, each of which is associated with a different level of awareness and a different degree of complexity in the way its detection and response systems are linked together.

- *Level 1*, the most basic form of processing, takes place automatically, completely outside of consciousness, by means of fixed cellular or neuronal pathways between the receptors that detect the information and the effectors that respond to it.
- *Level 2*, which is associated with the type of consciousness that sentient animals experience, is characterized by the integration of various sources of sensory and emotional information, so that the response to a given stimulus can vary, depending on the nature of this other input.
- *Level 3*, the most complex one, is associated with reflective consciousness and involves a conscious assessment of the detected information and of the options available for responding to it, based on the individual's accumulated knowledge and understanding. Responses at this level are determined by the meaning individuals give to the detected information, its significance to them, and its envisioned consequences, rather than by the information itself.

Plants, cells, microorganisms, and most invertebrates process information entirely by means of Level 1 pathways, mammals and other sentient creatures do so through a mixture of Level 1 and Level 2 connections, and humans use a combination of all three ways of linking detection and response together. While the responses to Level 1 processing are entirely predetermined, the responses to Level 2 show some variability, since they depend in part on the particular context. However, responses to Level 3 processing can vary a great deal from one individual to the next, since they result from choices that are shaped by personal beliefs and experience. The time delay between detection and response also differs between the three levels because of the different number of cellular and neuronal connections involved, with Level 1 taking the shortest time and Level 3 the longest. This can be seen when we accidentally touch a hot stove: first, we automatically pull our hand away (Level 1), then, a few hundredths of a second later, we experience the pain this has caused (Level 2), and finally, a few hundredths of a second after that, we become aware of what has happened and decide what to do about it (Level 3).

These different levels of information processing correspond roughly to the three evolution-based levels of brain development proposed by MacLean (1990). He maintains that the human brain is comprised of a primitive core that is similar to the brain of reptiles, over which is layered a set of structures similar to those in early mammals (which he calls the *limbic system*), and spread out over that is the most recently developed structure, the *neocortex*. The characteristic cellular architecture of the neocortex is present only in mammals and is most highly developed in humans. While not everyone agrees with the details of this particular formulation, the general idea of a layering of newer brain structures (and functions) over older ones has wide acceptance. The reptilian part of the brain, which includes the brain stem and cerebellum, controls basic homeostatic and survival mechanisms, such as the autonomic nervous system's regulation of breathing, heartbeat, and body temperature. Its regulatory activities consist of predetermined responses to particular information patterns, all of which normally take place automatically, outside of awareness (i.e., Level 1 processing). The limbic system, which includes the hippocampus, amygdala, and hypothalamus, is concerned with emotions and instinctive behaviors, including feeding, fighting, fleeing, and sexuality. It is responsible for generating the feeling states that enable us to determine the value (good or bad) of the input we receive from our external and internal senses, and for generating responses based on these appraisals (i.e., Level 2 processing). The neocortex, which comprises about two thirds of the human brain, is concerned with abstract thought, reasoning, language, decision making, and goal-directed behavior (i.e., Level 3 processing). Although developmentally distinct, these three brain regions are intricately interconnected, so that they function as one in processing different types of information. The general rule seems to be that the more complex an organism's information detection and response systems, the more elaborate the centralized structures (brains) required to coordinate them, and the greater the degree of conscious awareness.

Nonconscious Activity

Most of the information processing that takes place in the living world occurs silently and automatically through genetically based detection and response systems, both in species that are limited to this level of activity and in ones that are not. Biological entities that are confined to Level 1 processing are essentially stimulus bound and unable to change the way they respond to the information they detect. Although they may alter how they respond through learning, they have no choice but to utilize the new response once it is learned. The connections between the receptors that detect information and the effectors that respond to it automatically have been shaped by natural selection to meet the needs of the particular species. The biological processes that operate this way range from intracellular signaling, to the tropisms of single-cell organisms, to the intelligent behavior of invertebrates, such as octopuses and social insects. Intelligent behavior does not require consciousness, since it can result from a fixed sequence of automatic feedback responses. The female digger wasp, for instance, paralyses a cricket with its sting and then drags it to her burrow to serve as food for the eggs she has laid. She initially leaves the cricket outside the burrow while she enters to inspect it, and does not drag it inside until after this task has been completed. If the cricket is moved a few inches while the wasp is in the burrow, she will not drag it into the burrow until she has repeated her inspection routine. If the cricket is moved repeatedly, the wasp has to start over each time, apparently unable to escape from this genetically programmed sequence (Hofstadter 1979). The interplay of linked feedback mechanisms, in which the completion of one element acts as the signal that initiates the next one, is likely responsible for the complex instinctive behaviors seen in nonmammalian species, including the dance language of honeybees and the courtship rituals of certain birds.

A great deal of the information processing in our own brains also takes place outside of conscious awareness (Wilson 2002). We are, for instance, completely unaware of the physiological systems that regulate our bodily functions, or of the host of neuronal interactions that transform the patterns our retinal receptors detect into the images we see, or of how we choose the particular words we use when we speak or write. Our conscious mind is like the captain of an ocean-going liner that receives reports and issues commands from the bridge, while all of the hard work goes on below decks. Some of these nonconscious processes intrude into consciousness at times as Freudian slips or behavioral compulsions, like repetitively cleaning or checking. Nonconscious drives and motivations can also lead people to gratify needs and desires they are unaware of and may consciously disown. Behaviors that are shaped by nonconscious processes tend to be relatively fixed and inflexible, just as they are in the biological entities that can only process information this way. We are, for the most part, only aware of the objects and events that might have a perceptible impact on us or on which we might have a perceptible impact, but not the routine activities that regulate themselves (Zeman 2003, p. 269).

Choice

Consciousness is primarily about choice (Freeman 2001). Sentient consciousness, for instance, enables animals to choose the contextual information to include in determining how to respond to a perceived object or event. The way they respond to this information does not seem to be chosen consciously, however, but is based instead on preprogrammed algorithms. A simple model for making such a rule-based decision might look something like: if A and B are sensed, institute response X, while if A and C are sensed, institute response Y. A deer can choose its escape route from a predator not only by responding to the feelings the predator generates, but also by utilizing other types of sensed information, such as the direction of the wind, the availability of cover, and the location of nearby hiding places. Reflective consciousness, on the other hand, allows us access to a wider range of information sources and a wider selection of response options. We can, for instance, choose between externally or internally generated information, select whether to assess this through logical rules or intuitive feelings, and decide to respond in ways that advance our future goals rather than our present interests (Lombardo 2006).[6]

No one has any idea about how the brain actually makes conscious choices. There is some type of choosing process, but no *chooser*, although it certainly feels like there is one. Even if there were some sort of neuronal self that was making these choices, we would still be left to puzzle about what caused it to make the ones it did. Evolution is another process in which choices are made over time, even though there is no chooser; it just looks like it is a purposive endeavor in the same way that conscious choices do. The choices we make are, however, mainly determined by the type of information we consider in making them. How we respond to a given set of circumstances depends, among other things, on whether we consider the longer-term consequences of our actions, whether we think about the effect they have on others, and whether we care what people think of what we do.

Free Will

Although we believe we are free to choose what we do from among a number of options, we do not have any idea about how we do it. Even when we do not feel like it, we are able to make ourselves turn the other cheek, resist temptation, and save for a rainy day, simply by consciously exerting what we call *will power*. We can, for instance, will ourselves to get out of bed on a cold morning, even though we do not know exactly who or what is doing the willing. Purposive actions often involve a conscious choice between either gratifying a current desire or pursing a longer-term goal, a choice that animals that do not possess reflective consciousness are presumably unable to make. Children and adults with limited cognitive abilities also have less choice about how they respond, since they are not able to consider the more remote consequences of their actions. We distinguish our voluntary thoughts and actions from our involuntary ones by the sense that we have a choice about whether to initiate them, a sense that they are under our control.[7]

Reflective consciousness is the vehicle that enables us to choose what we do, rather than have our behavior determined by some type of preset algorithm. Free will is related to how we perceive and interpret the information we detect, as the meaning we give to it ordinarily determines how we respond. Since there is no objective way of deciding what perceived objects or events mean, we are free to decide this for ourselves—free to judge whether a neighbor is shy or rude, a friend principled or stubborn, or a politician flexible or wishy-washy—and to respond accordingly. Where we are most free, however, is in our ability to predict future events and set goals for ourselves, for these are mostly based on beliefs and expectations, not tangible realities. Reflective consciousness lets us imagine how things might turn out, which enables us to plan for the more distant future in ways that no other animal is able to do. We alone can generate goals for ourselves, develop strategies for attaining them, monitor our progress, and modify our plans when needed.

Our everyday lives and our social institutions are based on the belief that we have a degree of personal control over what we do, so that we are not just some type of elaborate machine whose responses are all predetermined. Some of our ancestors thought that our ideas and actions were animated by some type of *vital force*, an ethereal substance that left us when we died. Others believed that we were controlled by invisible forces emanating from the heavenly bodies or from various spirits and demons. These ideas lost favor over time as they were gradually replaced by more tangible versions of causation and responsibility. Ethics and morality are anchored in our sense of free will, since we can only be held accountable if we are free to choose what we do. We do not consider other animals to be morally responsible for their actions because we do not think they have free will, even though they may have a degree of choice about how they behave. Unfortunately, although the concept of free will explains effects that are otherwise inexplicable, it does little to help us understand them (Tye 1995).[8]

Consciousness remains an enigma. We have no idea how a mere wish can cause us to do something, how we can create thoughts and images that are independent of external cues, how future-oriented goals can cause us to forgo current pleasures, or how nonmaterial entities, like beliefs and aspirations, can affect material ones, like neurons and muscles. Dennett (1991, p. 33) is only partially correct, however, when he notes: "There is only one sort of stuff, namely matter—the physical stuff of physics, chemistry and physiology—and the mind is somehow nothing but a physical phenomenon," since the mind is not a material entity in the usual sense. It traffics in information, not matter and energy, even though it needs them to support its functions. As Searle (1994, p. 51) points out, the mind is a higher-order (emergent) feature of the brain that cannot be explained by its physical components, just like liquidity is a higher-order function of the way water molecules interact, not something the molecules themselves possess.

Notes

[1]Consciousness has become a topic of increased scientific interest over the past 20 years. The subject has attracted scholars from a variety of disciplines, including philosophers such as Dennett (1991), Chalmers (1996), McGinn (1999), Searle (2002), Honderich (2004) and Kriegel and Williford (2006); cognitive psychologists, like Ornstein (1991), Barrs (1997), Velmans (2000), and Gray (2004); mathematicians, like Penrose (1994); biologists, like Crick (1994) and Revonsuo

(2006); and neuroscientists, such as Greenfield (1995), Bennett (1997), Edelman and Tononi (2000), Zeman (2003), and Koch (2004). None of these authors deals with the information-processing aspects of consciousness.

[2]The slow alpha waves in an EEG that usually accompany these periods of relaxation probably reflect a diminution of cortical activity. Huxley (1954) and Castaneda (1968) provide literate accounts of the increased vividness of ordinary sensory experiences, the distortions of time, and the loss of self-awareness that certain mind-altering drugs induce.

[3]A number of ethologists and animal rights activists believe that other species share our sense of consciousness because of the complexity of their behaviors, especially their use of tools, problem-solving abilities, navigation skills, and capacity for social communication (Griffin 1976 and 1984; Walker 1983, and Ristau 1991). There is, however, no agreement about whether any of them share our reflective type of awareness. Call (2005, p. 321), for instance, argues that the great apes likely have some degree of self-awareness, as demonstrated by their ability to seek additional information in experimental situations of uncertainty, while Hauser (2001) maintains that they do not appear to have a sense of self or future goal-setting abilities.

[4]Our sense of our *self* as a unique individual in space and time, separate from the rest of the universe, and distinct from other beings, has at least three components: (a) a *representational self* comprised of a mental representation of what we look and sound like that allows us to recognize ourself in a photograph or on a voice recording, (b) a *conceptual self* consisting of a compilation of the ideas we have about our distinctive interests, tastes, talents, beliefs, and values, and (c) an *executive self* that perceives what is happening, chooses how to respond, and initiates independent thoughts and actions. Gallup (1970) reports that chimpanzees and human infants (but not monkeys) can recognize themselves in a mirror, as indicated by noticing a colored spot that had been painted on their forehead while they were asleep or anesthetized. While this indicates that they have a form of representational self, it does not mean they possess the other components.

[5]Cassell (1991, p. 36) notes: "Because of its temporal nature, suffering can frequently be relieved in the face of continuing distress by causing the sufferers to root themselves in the absolute present — 'this moment and only this moment.' Unfortunately, that is difficult to accomplish. It is noteworthy, however, that some Eastern theologies suggest that desire is the source of human suffering. To end suffering one must give up desire. To give up desire one must surrender the enchantment of the future."

[6]Choice is an aspect of consciousness that has largely been overlooked. It is not mentioned by any of the 22 experts in Blackmore (2006). James (1890, p. 8), however, observes: "The pursuance of future ends and the choice of means for their attainment are thus the mark and criterion of the presence of mentality in *a phenomenon.*" Mountcastle (1980) also notes: "The presence of the conscious control of action may, as an operational definition, be assumed when the organism displays the capacity for choice of action, the ability to set one goal aside in favor of another, the power to withhold action or reaction." Koch (2004, p. 319) sees choice as a way of determining whether an animal experiences conscious thought: "Force the organism to make a choice, such as inhibiting an instinctual behavior, following a delay of a few seconds. If the creature can do so without extensive learning, it must make use of a planning module that, at least in humans, is closely linked to consciousness."

[7]Free will implies that, although our choices are shaped by our past experiences and neuronal configurations, they are not automatically determined by them, since we could, supposedly, have behaved otherwise if we wanted to. Mele (2006) reviews current theories about free will.

[8]Jaynes (1976) maintains that our ancestors apparently had little subjective sense of self or free will prior to about 3,000 BC. The written records that have survived from the early Egyptian, Greek, and Mesopotamian civilizations portray people experiencing their own behavior as being controlled by outside forces, in the same way they thought these forces controlled everything else around them. They believed that human behavior was determined by the sun, the moon, and the stars, or by various spirits and deities. The notion of a reflective self does not begin to appear in the literature until about 1,000 BC, when men and women begin to be described as determining their own behavior. Although Jaynes' dating may not be entirely accurate, his general thesis is probably not far off the mark.

References

Austin JH (2006) Zen-Brain Reflections: Reviewing Recent Developments in Meditation and States of Consciousness. MIT Press, Cambridge

Barrs BJ (1997) In the Theater of Consciousness. Oxford University Press, New York

Bennett MR (1997) The Idea of Consciousness: Synapses of the Mind. Harwood, Amsterdam

Call J (2005) The self and other: A missing link in comparative social cognition. In HS Terrace, J Metcalfe (Eds), The Missing Link in Cognition: Origins of Self-Reflective Consciousness. Oxford University Press, New York

Cassell EJ (1991) The Nature of Suffering. Oxford University Press, New York

Castaneda C (1968) The Teachings of Don Juan: A Yaqui Way of Knowledge. University of California Press, Los Angeles

Chalmers D J (1996) The Conscious Mind. Oxford University Press, New York

Crick F (1994) The Astonishing Hypothesis. Simon & Schuster, New York

Crook JH (1980) The Evolution of Human Consciousness. Clarendon Press, Oxford

Csikszentmihalyi M (1990) Flow: The Psychology of Optimal Experience. Harper & Row, New York

Csikszentmihalyi M (1997) Finding Flow: The Psychology of Engagement with Everyday Life. Basic Books, New York

Blackmore S (2006) Conversations on Consciousness: What the Best Minds Think About the Brain, Free Will, and What it Means To Be Human. Oxford University Press, New York

Damasio AR (1999) The Feeling of What Happens: Body and Emotion in the Making of Consciousness. Harcourt Brace, New York

Damasio AR (2010) Self Comes to Mind: Constructing the Conscious Brain. Pantheon, New York

Dennett DC (1991) Consciousness Explained. Little Brown, Boston

Edelman GM (1992) Bright Air, Brilliant Fire. Basic Books, New York

Edelman GM, Tononi G (2000) A Universe of Consciousness: How Matter Becomes Imagination. Basic Books, New York

Freeman S, Herron JC (2007) Evolutionary Analysis, 4th edn. Benjamin Cummings, London

Freeman WJ (2001) How Brains Make Up Their Minds. Columbia University Press, New York

Gallup GG Jr (1970) Chimpanzees: Self-recognition. Science 167: 86–87

Gray J (2004) Consciousness: Creeping Up on the Hard Problem. Oxford University Press, Oxford

Greenfield SA (1995) Journey to the Centers of the Mind: Toward a Science of Consciousness. WH Freeman, New York

Greenfield S. (2000) The Private Life of the Brain: Emotions, Consciousness, and the Secrets of the Self. John Wiley & Sons, New York

Greenspan SI, Shanker SG (2004) The First Idea: How Symbols, Language, and Intelligence Evolved from Our Primate Ancestors and Modern Humans. Da Capo Press, Cambridge

Griffin DR (1976 The Question of Animal Awareness. Evolutionary Continuity of Mental Experience. Rockefeller University Press, New York

Griffin DR (1984) Animal Thinking. Harvard University Press, Cambridge

Hauser M (2001) Wild Minds: What Animals Really Think. Henry Holt, New York

Hofstadter DR (1979) Godel, Escher, Bach: An Eternal Golden Braid. Basic Books, New York

Honderich T (2004) On Consciousness. University of Pittsburgh Press, Pittsburgh

Huxley A (1954) The Doors of Perception. Harper & Brothers, London

James W (1890) Principles of Psychology. Henry Holt, New York

Jaynes J (1976) The Origins of Consciousness in the Breakdown of the Bicameral Mind. Houghton Mifflin, Boston

Kriegel U, Williford K (2006) Self-Representational Approaches to Consciousness. MIT Press, Cambridge

Koch C (2004) The Quest for Consciousness: A Neurobiological Approach. Roberts & Co, Englewood

Le Bon G (1903) The Crowd. Fisher Unwin, London

Lombardo T (2006) Development of the Human Capacity to Think about the Future. Author House, Bloomington

McGinn C (1999) The Mysterious Flame: Conscious Minds in a Material World. Basic Books, New York

MacLean PD (1990) The Triune Brain in Evolution: Role in Paleocerebral Functions. Plenum Press, New York

Mandler G (1997) Human Nature Explained. Oxford University Press, New York

Mayr E (1961) Cause and effect in biology: kinds of causes, predictability and teleology are reviewed by a practicing biologist. Science 134: 1501–1596

Mele AR (2006) Free will: Theories, analysis, and data. In S Pockett, WP Banks, S Gallagher (Eds), Does Consciousness Cause Behavior? MIT Press, Cambridge

Monod J (1971) Chance and Necessity. An Essay on the Natural Philosophy of Modern Biology. Alfred P Knopf, New York

Mountcastle VB (1980) Sleep, wakefulness, and the conscious state: Intrinsic regulatory mechanisms of the brain, In VB Mountcastle (Ed), Medical Physiology. CV Mosby, St. Louis

Ornstein R (1991) The Evolution of Consciousness. Touchstone, New York

Penrose R (1994) Shadows of the Mind. Oxford University Press, New York

Revonsuo A (2006) Inner Presence: Consciousness as a Biological Phenomenon. MIT Press, Cambridge

Ristau CA (1991) Cognitive Ethology: The Minds of Other Animals. Lawrence Erlbaum, Hillsdale

Searle JR (1994) Some relations between mind and brain. In RD Broadwell (Ed), Neuroscience, Memory, and Language. Library of Congress, Washington

Searle JR (2002) Consciousness and Language. Cambridge University Press, Cambridge

Terrace HS, Metcalfe J (2005) The Missing Link in Cognition: Origins of Self-Reflective Consciousness. Oxford University Press, New York

Tononi G, Edelman GM (1988) Consciousnesses and complexity. Science 282: 1846

Tye M (1995) Ten Problems of Consciousness: A Representation Theory of the Phenomenal Mind. MIT Press, Cambridge

Velmans M (2000) Understanding Consciousness. Routledge, London

Walker S (1983) Animal Thought. Routledge & Kegan Paul, London

Wilson TD (2002) Strangers to Ourselves: Discovering the Adaptive Unconscious. Harvard University Press, Cambridge

Zeman A (2003) Consciousness: A User's Guide. Yale University Press, New Haven

Chapter 16
Maladaptive Behavior

Abstract One of the great marvels of evolution is how finely matched every species is to the ecosystem in which it lives, for natural selection weeds out the genes of its less well-adapted members. *Homo sapiens* is an anomaly in this grand scheme of things, since we are prone to a wide range of maladaptive behaviors that diminish our well-being and survival, including consuming addictive substances, behaving irresponsibly, and ingesting more calories than we need. Although evolution has provided us with a brain that is able to use the information it detects to build models of how the world works, there is no mechanism for assuring that the information is correct or that the models are accurate. Most of our maladaptive behaviors are due to defects in the way we create symbolic representations, interpret meaning, and set goals for ourselves, since these processes are all subject to error.

One of the great marvels of evolution is how finely matched every species is to the ecosystem in which it lives. This is because the organisms and their biologic environments have evolved together, with changes in one leading to changes in the other. It is why nocturnal bats developed sonar, why the introduction of antibiotics led to the emergence of resistant bacteria, and why the different beak shapes of Darwin's finches match so closely the different foods they eat (Sulloway 1982). Nothing evolves in isolation; every living thing is part of an ecological system in which all the various animals, plants, and organisms interact and regulate each other. Everything in the system is kept in balance; perturb one element and the others all respond. Natural selection sculpts each species to fit the particular niche in which it lives by weeding out the genes of the individuals who do not function effectively in it.

Homo sapiens is, in many ways, an anomaly in this grand scheme of things. Despite being blessed with the widest array of talents for detecting and responding to information, humans frequently misuse these gifts and behave in grossly maladaptive ways. The great irony is that, having developed a set of skills that enables us to be relatively safe from natural perils, we now use them to harm ourselves. No other species indulges in the sort of self-injurious behaviors that we do. None of

A. Reading, *Meaningful Information: The Bridge Between Biology, Brain, and Behavior*,   135
SpringerBriefs in Biology 1, DOI 10.1007/978-1-4614-0158-2_16,
© Springer Science+Business Media, LLC 2011

them overeats, gets out of shape, deliberately ingests or inhales toxic substances, or routinely maims or kills other members of its own kind. Although other animals may not always act in ways that are in their own best interest, it is not in their nature to injure themselves. But, somehow, it is in ours. We are prone to a wide range of maladaptive behaviors that diminish our own quality of life and survival, including consuming addictive substances, being indolent or negligent, and ingesting more calories than we need. We also engage in behaviors that endanger other members of our species and threaten the rest of the planet by polluting the environment, mismanaging ecosystems, and manufacturing weapons of mass destruction.

Flawed Gifts

Natural selection produces adaptations that are good enough for the purpose at hand, but not necessarily ideal. It is based on favoring genetic changes that increase an organism's reproductive fitness, not on perfecting the various species. The process can tolerate a certain amount of error as long as it does not interfere with an organism's ability to pass its genes along to the following generations. Thus, although evolution has provided us with a brain that is able to use the information we detect to build models of how the world works, it has left the scope and nature of that information up to what each of us chances to encounter. The models we develop are thus error prone since they are shaped by the culture in which we happen to be raised and the particulars of our individual experience. A brain that is lucky enough to develop in a structured and nurturing environment constructs a different model of the world than one that develops in a chaotic and uncaring one. The major flaw in our information-processing system is the ease with which erroneous beliefs and ideas can be incorporated into our mental models and distort the abilities that depend on them, such as setting achievable goals. Flawed models lead people to misperceive and misinterpret the information they detect, which can cause them to respond in ways that are not in their own best interest—for we respond to how we perceive and interpret events, not to the events themselves.[1]

Early experiences have a profound effect on the way we later see the world, since the models we construct during our formative years become the reference standards against which we evaluate later information. We understand the meaning of the objects and events we encounter by appraising them against the mental models we have developed and misunderstand them and respond inappropriately when our models fail to portray events and relationships accurately. Individuals who are brought up in settings affected by violence, drunkenness, abuse, or neglect, for instance, develop distorted models of how the world works, since these are all toxic to the growing mind. Children need to be raised in a supportive and predictable environment for their information-processing capabilities to fully develop—in much the same way that seeds need a favorable combination of soil, sun, and water to blossom into flowers, even though they contain all of the necessary genetic information. The dilemma we all face is that, although we are endowed with an unrivaled ability to learn from what we experience, there is no guarantee that what we learn is correct.

For most of us, the flaws that creep into how we model the world do not usually cause much trouble, just an occasional lapse in judgment, an unwise investment, or a behavioral indiscretion. But, the amount of maladaptive behavior can be so great in some individuals that it disrupts their ability to enjoy satisfying and productive lives. While mental illnesses constitute the more extreme forms of maladaptive functioning, they are not the only ones. The catalog includes a wide variety of other self-harming activities, including drug and alcohol abuse, cigarette smoking, gluttony, sexual deviance, personal neglect, and pathological levels of traits like suspiciousness, procrastination, self-absorption, and thrill-seeking. We can also behave in ways that injure others through exploitation, abuse, bigotry, violence, and crime, although these behaviors are not necessarily maladaptive from the perpetrator's point of view.[2]

Our worldview is the prism through which we experience everything else, the framework we use to interpret the meaning of the objects and events we encounter. Since the meaning we give to events determines how we respond to them, much of our behavior is shaped by the way we come to understand the world around us. Individuals from similar backgrounds usually share many of the same beliefs and values, and these bind them together in everything from gangs and social groups to nations and terrorist organizations. Differing beliefs and values, on the other hand, can separate people just as easily, both within and between particular social or political entities. We tend to associate with people whose views are similar to our own, avoid ones with different beliefs, listen to pundits who echo our own sentiments, and read books and articles that endorse our particular convictions. Unfortunately, the more our ideas are reinforced in these ways, the more we come to believe they are true and that contrary ones are not. We are an odd species for, while other animals fight over territory or mates, we fight over beliefs and ideologies—over what are, in effect, just alternative ways of interpreting the information we detect.

Erroneous Information

The unique information-processing gifts that evolution has bestowed on us come with a price, for most of our maladaptive behaviors are due to flaws in the way we create symbolic representations, build mental models, interpret meaning, and set goals for ourselves. There is, unfortunately, no mechanism for assuring that the models and representations we build are accurate, the meanings we infer are correct, or the goals we set are realistic. Our models of how the world works are constructed from what we personally experience and what other people tell us, but there is no guarantee that our perceptions are correct or that what others pass on to us is true. We have no internal *truth meter*, no independent way of discerning whether what we are told, perceive, or infer is correct. All our brain can do is determine whether incoming information seems reasonable and consistent with what we already know, not whether it is really true. False beliefs that fit with how we understand the world are not readily altered by factually correct information, since this can easily be discounted or denied (Brafman and Brafman 2008). Most people do not even consider the possibility that it is their beliefs that may be wrong when what they see or hear contradicts them.[3]

Our lives are shaped by the decisions we make, both consciously and otherwise; by which paths we choose to follow and which ones we decide to avoid. Rather than having relatively fixed ways of responding to what we detect, we are confronted instead with a myriad of choices about what we should do, and where, when, and with whom we should do it. Should I get married, ask for a raise, go to college, try cocaine, buy a new TV, visit my grandmother, or learn the guitar? From a biological point of view, rational decisions are ones that utilize our resources efficiently, while irrational choices are ones that do not. Logical decisions are thus not always rational within this framework, since they can result in maladaptive behavior when they are based on false premises, and rational decisions are not always logical, since they can be based on intuitive appraisals. Thus, while evolution has enabled us to make more choices than any other species, it also has allowed us to make more mistakes.[4]

Inaccurate Predictions

Some of the decisions we make are relatively trivial because they do not affect our overall well-being, like choosing between cornflakes and Rice Krispies for breakfast. Others can have a major impact on our lives, either positively or negatively, like deciding to start our own business, run for public office, or quit smoking. Most of the more important decisions we make have long-term consequences, so that the more clearly we can predict these, the more likely we are to make correct choices. We peer into the future to set goals for ourselves and assess whether our current choices are likely to help or hinder attaining them. Goal-directed behavior involves determining the benefit of a desired future, the strategy needed to achieve it, the costs involved, and the likelihood of realizing it. While our brains compute these cost–benefit assessments almost instantaneously, largely outside of awareness, the predictions we make are only as good as the data on which they are based. No matter how confident we feel about our decisions about which course to follow, which beliefs to espouse, and which battles to fight, the hopes that flawed models generate typically lead only to frustration and disappointment.

The future is always somewhat of a gamble, with the challenge being to develop strategies that maximize the chance of success while minimizing the risk of harm (Frank 1961). Although we can usually make reasonably accurate predictions about commonly occurring events, we have trouble predicting outcomes when the situations are unfamiliar, the consequences are remote, or the available data are not consistent. Predicting what will happen is much like forecasting the weather, since it involves making an appraisal of the likelihood of certain events taking place, based on prior knowledge and experience. Because the future is always veiled in some degree of uncertainty, the best we can do is predict the probability of future occurrences, like the announcer who forecasts a seventy percent chance of rain. The more our models are flawed, however, the less accurate the predictions we base on them, and the greater our chance of being lead astray. Our predictions can go wrong either because an event we predicted does not occur (a false alarm) or an event we did not predict does occur (a surprise). Unfortunately, there is a trade-off between these two types of error, since the steps taken to reduce one inevitably increase the other.

Mental Illnesses

Mental illnesses are characterized by amounts of maladaptive behavior, erroneous decisions, and inaccurate predictions that are large enough to disrupt the person's ability to successfully cope with the challenges of everyday life, either on a temporary or a more permanent basis. Affected individuals tend to apply special meanings to various objects and events and respond accordingly, either because their brains have not developed appropriate information-processing connections or have formed inappropriate ones (Laing 1982). Especially in the more severe conditions, like schizophrenia and bipolar disorder, genetic abnormalities appear to disrupt the person's capacity to process certain types of information, and these difficulties are then exacerbated by stressful life experiences. The extremes of emotion that mentally ill individuals typically exhibit are probably just an indication of how stressful the routine tasks of living are for them, since their emotional states are usually in keeping with how they experience the world even though they may not make sense to outside observers.

Although our current diagnostic criteria enable us to classify these illnesses and identify ways to treat them, they do not tell us anything about what causes them or in what way the brains of affected individuals are abnormal. We do not know, for instance, whether the changes in neurotransmitter functioning that accompany many of these illnesses are related to the underlying abnormalities that cause them or to the brain's attempt to counteract their disruptive effects. One of the reasons the search for the biological causes of mental illnesses has been so difficult is that we do not know exactly what we are looking for, since we still do not know enough about how the normal brain functions (Brüne 2008). Although almost everyone acknowledges that disorders of the mind reflect disorders of the brain, there is no agreement about how the affected brains are different or what brings these differences about. Since true mental illnesses only seem to occur in human beings, the logical place to look for the abnormalities that cause them is in the areas of the brain that are most distinctively human, like the prefrontal cortex. What we will most likely find, however, are differences in the way neurons are arranged and interconnected in these areas, rather than actual structural change (Insel 2010).[5]

Because mental illnesses represent disruptions of the neural circuits that ordinarily regulate our behavior and make it appropriate to the circumstances we encounter, many of their symptoms reflect flaws in the way affected individuals detect and respond to information. As McGuire and Troisi (1998, p. 118) point out, "To discuss mental conditions is also to discuss information recognition and information signaling." Hallucinations, for instance, are due to abnormal perceptions that fail to distinguish between internal and external sources of information. Delusions are erroneous beliefs that result from misinterpreting the meaning of detected information, so that paranoid individuals see threats and conspiracies where none really exists. Children with attention deficit disorder cannot suppress inadvertent sensory stimuli, and ones with autism are unable to interpret the meaning of social cues (Hobson 2004). Anxiety disorders are associated with unwarranted fears that something bad will happen, as in the way that fearful meanings become attached to nonthreatening objects in the various phobias. Grief and depression, on the other hand, are indications that something has gone amiss with the individual's expected future, usually as the result of a real or imagined loss.

Although the symptoms of most mental illnesses fluctuate a good deal over time, the crucial brain defects that produce them presumably remain relatively unchanged. If we want to discover the underlying causes of these disorders, we need to be able to identify brain dysfunctions that are present when affected individuals are in remission, not just when they are acutely ill. There is evidence, for instance, that significant amounts of cognitive dysfunction are still present in schizophrenic patients who are not currently symptomatic, including abnormalities in verbal learning, memory, psychomotor speed, and vigilance tasks (Schretlen 2007). Many of these patients also continue to have a residual amount of thought disorder when they are in remission, including some looseness of associations and difficulty in understanding abstractions. However, no one has yet been able to identify a core set of cognitive deficits that is present in everyone with the disorder. Either we have not looked the right way to find them or, more likely, schizophrenia is not a single pathological entity but is the final common pathway through which a number of major information-processing dysfunctions get expressed.

One of the most puzzling aspects of mental illnesses is why they have not been eliminated from our species during the course of evolution, since natural selection operates by weeding out the less fit members of a species. There are at least two possible reasons for this. One is simply that the genes that predispose us to these disorders also have offsetting effects that benefit us, in much the same way that the genes for sickle cell anemia protect us from malaria. Some looseness of associations or unusual thought process, for instance, might contribute to creativity and discovery. A more likely explanation, however, is that these various forms of maladaptive behavior are unavoidable consequences of the unique way we process information and cannot be separated from them. They are like the side effects that predisposed individuals experience when they take certain medications. The mental disruptions that susceptible individuals experience would then be considered to be untoward "side effects" of their information-processing capabilities.

Our genes provide us with the means for processing complex forms of information, but do not determine the information we actually process, like the way they provide us with the capacity for language, but not the actual language we speak. The circuitry in the neocortex is largely shaped by how we experience the world, so that aberrant experiences lead to aberrant brain functioning, much as they do in traumatic stress disorders. Small genetic differences in any of the abilities involved in processing information, constructing models, and making decisions could, under such a scenario, be greatly amplified by certain types of adverse experience. These genetic differences could then pass from generation to generation without having a noticeable effect on individuals who do not have these types of adverse experience, since the genes would not be sufficient to cause the disorders on their own.

Mental illnesses are, in effect, part of the price we pay for having such special brains. They represent malfunctions of some of the unique information-processing abilities our brains possess, such as abstract thinking, imagining, anticipating, and reflecting. Although other animals show disturbed and maladaptive behavior when they are stressed, these reactions are not true replicas of human mental illnesses. Our understanding of mental disorders has, in fact, lagged behind our understanding of physical ones because there are no naturally occurring animal models that can be used to investigate them. Ironically, although mentally ill individuals have often been stigmatized by being considered less than fully human, their illnesses actually affirm their humanity.

Notes

[1] Linden (2007) describes the human brain as "a cobbled-together mess" in which newer structures and newer functions have just been layered over older ones as a result of evolution, rather than being "designed" to fit together or replace them.

[2] Within-species violence can result in a reproductive advantage for the aggressors by eliminating potential competitors. Genes that predispose to such behavior can thus become increasingly represented in a population if it has no countervailing measures to limit them (Wrangham and Peterson 1996). Most cultures have developed social, moral, and legal sanctions for controlling aggression within social and ethnic groups, but not necessarily between them.

[3] Science represents a systematic way of understanding how the universe functions. It differs from our individual ways of determining truth and falsity in that it endeavors to gain a broad consensus about its findings by performing controlled experiments, having others replicate and verify its results, reducing subjective biases, and stating its beliefs and assumptions explicitly. Scientists arrive at the truth by systematically identifying things that are untrue and building theories and models based on facts and ideas that have survived the attempts to refute them.

[4] Behavioral economics is a relatively new discipline that studies the psychological factors that bias logical decision making and examines why markets are not always predictable on the basis of rational expectations (Kahneman and Tversky 2000). In addition to the distortions caused by faults in our individual mental models, there are a number of built-in information-processing biases that affect the decisions all of us make. These include tendencies to be averse to loss, favor the immediate over the deferred, and overestimate our own abilities. Most choices are affected by the way the options are presented, so that the items purchased at a supermarket are influenced by where they are placed on the shelves (Thaler and Sunstein 2008). We also tend to follow others rather than rely on our own judgment, especially when they seem to know what they are doing, since we mistakenly equate the strength of their convictions with how correct they are. Marcus (2008, p. 60) describes various cognitive errors that interfere with our ability to think rationally and distort the decisions we make. As he observes, "While evolution gave us the gift of deliberate reasoning, it lacked the vision to make sure we used it wisely."

[5] The prefrontal cortex is a part of the brain that is significantly more developed in humans than in other primates. It is located at the front of the brain, where it causes our foreheads to bulge outward rather than slant backwards, like those of our primate cousins and hominid ancestors (Preuss 2000). It plays a central role in orchestrating our executive and decision-making abilities, including planning for the future and choosing between different courses of action (Struss and Benson 1986; Fuster 1997). Patients with gross damage to this area are typically unable to initiate future-directed behavior and, as a result, spend their days mostly in the present, without ambition, foresight, or spontaneity (Bechra et al. 1994).

References

Bechra A, Damasio A, Damasio H, Anderson S (1994) Insensitivity to future consequences following damage to human prefrontal cortex. Cognition 50: 7–15

Brafman O, Brafman R (2008) Sway: The Irresistible Pull of Irrational Behavior. Doubleday, New York

Brüne M (2008) Textbook of Evolutionary Psychiatry: The Origins of Psychopathology. Oxford University Press, New York

Frank J (1961) Persuasion and Healing. Johns Hopkins University Press, Baltimore

Fuster JM (1997) The Prefrontal Cortex, 3rd edn. Lippincott-Raven, Philadelphia

Hobson P (2004) The Cradle of Thought: Exploring the Origins of Thinking. Oxford University Press, New York

Insel T (2010) Understanding Mental Disorders as Circuit Disorders. Cerebrum, February Issue. The Dana Foundation, New York

Kahneman D, Tversky A (2000) Choices, Values, and Frames. Cambridge University Press, Cambridge

Laing RD (1982) The Voice of Experience. Pantheon, New York

Linden DJ (2007) The Accidental Mind. Harvard University Press, Cambridge

McGuire M, Troisi A (1998) Darwinian Psychiatry. Oxford University Press, New York

Marcus G (2008) Kludge: The Haphazard Construction of the Human Mind. Houghton Mifflin, New York

Preuss TM (2000) What's human about human behavior? In MS Gazzaninga (Ed), The New Cognitive Neurosciences, 2nd edn. MIT Press, Cambridge

Schretlen DJ (2007) The nature and significance of cognitive impairment in schizophrenia. Johns Hopkins Advanced Studies in Medicine 7(3): 72–78

Struss DT, Benson DF (1986) The Frontal Lobes. Raven Press, New York

Sulloway FJ (1982) Darwin and his finches: the evolution of a legend. J. Hist. Biol. 15:1–53

Thaler RH, Sunstein CR (2008) Nudge: Improving Decisions About Health, Wealth and Happiness. Yale University Press, New Haven

Wrangham R, Peterson D (1996) Demonic Males: Apes and the Origins of Human Violence. Mariner Books, New York

Chapter 17
Fabricated Devices

Abstract Given the vast array of fabricated devices that perform information-related tasks in our culture, it is no wonder that the biological aspects of information often get overlooked. Virtually all of the fabricated devices that have ever been invented represent ways of extending our biological abilities for detecting, processing, or responding to meaningful information. But they are unable to do any of this on their own, apart from some form of human intervention. They are also unable to process emotional information, so they have no appreciation of what they are doing or why they are doing it. Information Theory is not really a theory about information, at least not the meaningful kind that helps regulate living entities. It deals with the capacity of a system to transmit information, rather than the meaning of the information being transmitted, which is why attempts to apply it in biology have not been particularly fruitful.

Given the vast array of fabricated devices that perform information-related tasks in our culture, it is no wonder that the biological aspects of information often get overlooked. But none of these devices can detect or respond to information on its own, apart from some form of human intervention. Nor can any of them distinguish meaningful from meaningless information without being programmed by someone to do so. All of these devices are, one way or another, simply instruments that expand our own biological information detection and response abilities. Modern science and technology have, however, only been made possible by the invention of devices that have been increasingly able to detect, store, transmit, and respond to meaningful information, for these are the tools of exploration and discovery. Although our amazing collection of electronic gadgets and computers may seem to challenge the idea that detecting and responding to meaningful information are primarily biological functions, none of these devices can operate on its own, independent of living beings who tell it what to detect and how to respond.

We live today in what has been called the Information Age, surrounded by machines that are capable of detecting, storing, transmitting, and processing information on a

A. Reading, *Meaningful Information: The Bridge Between Biology, Brain, and Behavior*,   143
SpringerBriefs in Biology 1, DOI 10.1007/978-1-4614-0158-2_17,
© Springer Science+Business Media, LLC 2011

previously unimaginable scale. But not all of the transmitted patterns of matter and energy represent meaningful information, since many of them fail to generate a response in the recipients. A great deal of what is referred to today as "information" in the media and on the Internet is, in fact, only data (patterns that are potentially meaningful) or noise (patterns that are not meaningful). Some of the content is too technical to be easily understood, some is encrypted to prevent unwanted detection, and some is in foreign languages, but a lot of it is simply inconsequential or irrelevant as far as most individuals are concerned. We are inundated with visual and auditory stimuli that are designed to provide us with information about what to buy, what to believe, and who to vote for, although most of it has little real effect on us.

Extenders

Most of the fabricated devices our species has invented represent ways of extending our ability to detect, process or respond to meaningful information. Our early ancestors developed stone tools to increase the ways they could respond to what they detected, and later invented wheeled vehicles, introduced agriculture, and used animals to supplement their own effort.[1] The Renaissance saw the emergence of the microscope and the telescope, both of which dramatically extended our ability to detect meaningful patterns of visual information. The modern scientific era then ushered in an avalanche of technological devices for detecting, storing, and transmitting information. These include the *scopes* that allow us visualize inaccessible objects (e.g., sigmoidoscopes and oscilloscopes), the *meters* that let us to measure hidden changes (e.g., tachometers and barometers), the devices that extend what we can detect (e.g., sonar and radar), and the telegraphs, telephones, and televisions that allow us to transmit information to distant places. We also have developed a multitude of devices that enhance our ability to respond to the information we detect, including the appliances we use, the automobiles we drive, the medications we take, the machines that make the products we buy, and the rockets and bombs we use to protect ourselves.

Intelligent Machines

Intelligent machines are devices that can both detect and respond to meaningful information. They differ from mechanisms that can only perform one of these functions, as well as from mechanical ones that respond primarily to energy, rather than information. The responses generated by intelligent machines are powered by energy supplied by the responding entity, not the initiating one, just as they are in biological systems. In most cases, the information these machines detect is in the form of on/off signals that control how they operate. A vending machine, for instance, delivers a soda or a candy bar when its sensors detect that the correct amount of change has been inserted. The motor that delivers the soda is, however, powered by externally supplied electricity, not by the signal it receives. A cruise missile is a more complex device that keeps on course by continually comparing what it senses against an internal map of the territory over which it is flying, and makes appropriate adjustments to its motors and control surfaces on this basis.

Intelligent behavior involves responding to detected information in ways that enable the entity to function effectively. It does not need to involve learning, understanding, or consciousness, since even simple unicellular organisms can behave intelligently, even though they possess neither a brain nor a nervous system. If the way *E. coli* moves toward detected nutrients and away from detected toxins is considered to be intelligent behavior, then machines that display comparable behavior also should. The main difference between the two is that the one has been programmed by an engineer and the other by natural selection. Intelligence in humans, however, extends far beyond the way machines can behave, since even the most sophisticated devices are not able to take advantage of fortuitous circumstances, make sense out of ambiguous or contradictory messages, or synthesize new concepts by putting old ones together in new ways (Hofstadter 1979, p. 26).

Many of today's intelligent machines utilize a combination of mechanical energy and informational controls, with the former being used to drive a particular process and the latter to regulate how it does it, like the way mini-computers are used to control today's automobile engines. The cybernetic systems described in Chap. 2 are the prototype of intelligent machines, since they use feedback to control their own operations. Although the first feedback mechanisms were mechanical in nature, like the governors that regulated James Watt's steam engines, these have largely been replaced by informational ones that utilize electronic signals to keep automated machines on target. The advent of digital computers and robotic devices has, however, taken the notion of intelligent machines to a completely new level. While cybernetic devices use informational signals to regulate how they operate, computers deal with logical reasoning and complex messages in ways that enable them to learn, remember, and make decisions. Robotic machines, on the other hand, are able to use input from sensors to detect information patterns in their environment, analyze whether they are meaningful to them, and then respond by initiating a procedure, manipulating an object, or moving in a particular way.

Artificial intelligence (AI) is a branch of computer science that uses complex algorithms to enable machines to perform information-processing functions that mirror our own intellectual abilities. The field has two major thrusts: (a) devising and studying models that simulate aspects of the neuronal mechanisms involved in perception, cognition, learning, and movement and (b) developing expert systems for analyzing complex data sets or implementing robotic controls, as in medical diagnostics and manufacturing. Artificial neural nets are computer algorithms that are able to learn by changing the strength of their connections in response to the input that gets fed into them, so that they do not need to be programmed to solve problems the way serial computers do (Forbes 2004). Because they are better at pattern recognition than serial computers, neural nets have been successfully used in face, handwriting, and voice recognition, even though they function more slowly than other computer operations.[2]

Humans Versus Machines

Many people thought that computers had finally become smarter than people when the IBM supercomputer Deep Blue defeated Garry Kasparov, the world chess champion, in a six-game match in 1997. Deep Blue, however, was a specific-purpose machine that was designed expressly to play chess, so that it could not engage in other tasks, had no other form of intelligent behavior, and neither understood nor cared about what it was doing. Kasparov, on the other hand, knew who and where he was, understood what was happening, and could turn his thoughts to other matters. Although brains and computers both process information, they do it in different ways and utilize different skills. Brains are better at recognizing verbal and visual patterns, differentiating shape from shading, and deriving three-dimensional structures from two-dimensional images, while computers are better at solving complex problems whose solution can be reached following a formalized series of instructions (Glynn 1999). Computers are better at solving complex problems and handling massive amounts of data, while brains are better at dealing with ambiguous situations and interacting with the environment (Searle 2004, p. 91). As Hawkins (2007) points out, even a 5-year-old child can understand spoken language, distinguish a cat from a dog, and play a game of catch far better than any computer.

Digital computers are essentially logic machines that process symbolic forms of information in an extremely rapid fashion according to built-in rules and algorithms (Butler et al. 1998). Brains, on the other hand, process both sensory and symbolic information, interpret its meaning, understand relationships, and initiate goal-directed behavior. One of the major differences between brains and computers, however, is that computers cannot process emotional information (Kelley 2005). As outlined in Chap. 13, what we experience as *feelings* is information about the meaning and value of the patterns of energy and matter we detect. Computers thus have no awareness of what the information they process means, whether or not it is correct, or whether it might harm someone, since the patterns of energy and matter they manipulate are essentially value-free. The fact that computers just deal with facts without being swayed by emotion is one of their principal virtues, but it is also one of their major shortcomings, since it deprives them of the value-based types of information that enables brains to choose how and when to respond. According to Picard (1997), computers will need the ability to recognize and express emotions in order to be genuinely intelligent and interact naturally with humans.[3]

Advocates of the so-called *computational theory of mind* tend to see the brain simply as a computing device. Montague (2006, p. 24), for instance, claims that "even our thoughts are equivalent to computational steps, only running on a very specific, biological evolved device: our brains" and that "the brain possesses all of the characteristics of a highly efficient computational machine." It is not clear, however, whether all of the information processing that goes on in the brain should be regarded as *computation*. Most dictionaries define computation as a process that involves the manipulation of mathematical symbols by physical systems, indicating that it refers to digital processes rather than analog ones. The brain is not limited to processing digital symbols, but can also use analogic evaluations to infer meaning, generate feelings, and make value-based choices. As Cairns-Smith (1996, p. 114) notes: "The neuronal computers in organisms were never really there to do arithmetic, but to serve as controllers. The ancestor of the brain was more like a thermostat than an abacus." The skill and complexity of intelligent devices are, in fact, a manifestation of the intelligence of their creators, not the devices themselves.[4]

Information Theory

Information Theory is a set of mathematical procedures developed by Claude Shannon for quantifying the amount of data that can be transmitted by electronic communication devices (Shannon and Weaver 1964). It revolutionized the way engineers designed communication systems and was one of the breakthroughs that led to the development of digital computers. It caused a considerable amount of confusion, however, when it was applied to biological systems. The problem is that Information Theory is not really a theory about information, at least not the meaningful sort that helps regulate living entities. The theory deals with the *capacity* of a system to transmit information, not with the content or meaning of the information being transmitted. Even though Shannon explicitly stated that he was using the word "information" in a special sense that should not be confused with its ordinary usage, this caution was largely forgotten as its influence became more widespread (Kåhre 2002). As Young (1987, p. 8) notes: "The Shannon formula is a measuring device, and so, to equate it with information in its general sense is to confuse a measuring device with what it measures. A formula that measures the amount of apples in a barrel obviously is not the same as the apples."

Shannon's theory defined information as something that is able to reduce uncertainty, and measured it by the amount of uncertainty it reduced. The amount of information in this model is like the number of yes and no questions needed to find the answer in a Twenty Questions game. The process involves a binary search algorithm, where every question is equivalent to one *bit* (binary unit) of Shannon-type information. The theory is essentially an engineering paradigm used to design systems for transmitting messages over a communication channel between a source and a receiver, with the capacity of the channel being measured in bits per second. Shannon's concept of information is best understood by considering the concept of uncertainty, so that the greater the amount of uncertainty, the greater the amount of information needed to resolve it. The amount of information in the ordinary sense of the word is related instead to the amount of organization and knowledge.[5]

Information Theory has relatively little relevance in living organisms, because it is the meaning of information that determines its biological effects, not the amount. Most biological information is transmitted as simple on/off signals that, even though they only carry the minimum possible amount of Shannon-type information (i.e., one bit), are able to trigger large and varied responses. As Von Bayer (2004) notes: "The trouble with Shannon's nifty operational definition is that it says nothing about the intended meaning of the message." As previously mentioned, two completely different messages, one loaded with meaning and the other with complete nonsense, can thus contain exactly the same amount of Shannon-type information (Severin and Tankard 1979). This is why the attempts to apply the theory to the nervous system or utilize it in molecular biology have met with only limited success (Sarkar 2005, p. 195). As Bolton and Hill (2004, p. 10) observe: "The mathematical theory is of no immediate relevance in explicating the concept of semantic (meaningful) information that underpins theorizing in cognitive psychology, and cognitive explanations of behavior in particular."

The major application of Information Theory has been in determining how to communicate error-free data through a noisy channel (Kosko 2006). The universe is a very noisy place, full of stray forms of energy that can interfere with information transmission systems, even though we cannot ordinarily detect them. Extraneous stimuli that hinder the detection of meaningful information constitute *noise* as far as the potential recipients are concerned. This is a particular problem in digital transmission systems where unrelated sources of energy, such as power surges, atmospheric radiation, and the "crosstalk" of other signals passing through the same channel, can disrupt the information being conveyed by changing some of their "0" bits into "1" bits, and vice versa. Stray radiation energy, either from the atmosphere or X-Ray machines, can also alter the meaning of genetic information by causing mutations that change the structure of one of the nucleotides on a strand of DNA. Since some amount of noise exists in all electronic devices as a result of thermal energy, the theory has been a key factor in the development of reliable communication systems. It has also been used to investigate the efficiency of neuronal signaling systems, although not the content (Baddeley et al. 2000). Signal-to-noise ratio is a construct for assessing transmission by comparing the level of a desired signal to the level of background noise. Engineers try to maximize signal-to-noise rations by filtering out extraneous input and increasing the power with which the message signal is sent. Some amount of built-in redundancy is thus helpful in transmitting error-free messages, which is probably why it is a built-in feature of most languages.

Information Theory's use of the term *entropy* as a measure of uncertainty, rather than of randomness and disorder, has also been source of confusion. Shannon called his uncertainty concept *entropy* because of its computational similarity to thermodynamic entropy but the two are not the same (Pierce 1980, p. 3). Rather than being a measure of disorganization, entropy in Information Theory is a measure of the amount of information conveyed by a message, based on the amount of uncertainty it reduces in the recipient (Küppers 1990). Thus, while entropy is inversely related to meaningful information (the greater the amount of information, the less the entropy), it is directly related to Shannon-type information (the greater the amount of information, the greater the entropy). As Wicken (1987, p. 23) observes: "As a result of its independent lines of development in thermodynamics and communication theory, there are in science today two 'entropies.' This is one too many."[6]

Notes

[1] The invention of the steam engine launched the Industrial Revolution by providing a source of mechanical power that no longer depended on the uncertainties of wind, water, or muscular strength. Steam was replaced by fossil fuels, electric power systems, and nuclear reactors, which then expanded our response abilities even further.

[2] Von Neumann (1958) describes how the idea that the brain could function like a digital computer was taken up by the emerging field of artificial intelligence (AI). Dreyfus (1972) challenged the euphoria that accompanied the early claims of the AI movement and predicted the difficulties that would eventually limit its accomplishments. As Edelman (2004, p. 5) notes: "Computer models of the brain and mind assume that the brain has a set of programs which are capable of changing states based on the information carried by the inputs, yielding functionally appropriate outputs. These models do not deal with the fact that inputs to the brain are not unambiguous — the world is not like a piece of tape with a fixed sequence of symbols for the brain to read."

[3] Alan Turing, one of the founders of the computer age, proposed a test to determine whether a computer could think (Turing 1950). The Turing Test involves a judge communicating by text messages with a human and a computer, both located in separate rooms away from the judge. Turing claims that the machine passes the test of being able to think intelligently if the judge cannot reliably identify which one he or she is dealing with. Detractors claim that this only shows that the machine can *act* like it is thinking, but does not prove this is actually what it is doing. The test is limited, however, because it only deals with computable symbols and does not include any of the other types of sensory inputs that people process. Even so, apparently no machine has yet passed it (Gray 2004, p. 126).

[4] Digital computers can be programmed to make analog-type comparisons between different sets of binary data. This is the basis of the *search* and *spell-check* algorithms on desktop machines, as well as the complex programs used to match nucleotide sequences to known genome configurations in bioinformatics. Digital computer comparisons only work with information patterns that can be converted into the binary code sequences. The analogic matching of incoming sensory information with stored templates seems to be based on comparing the patterns of activated neurons with reference patterns, rather than on a digital code.

[5] Schneider & Sagan (2006, p. 23) note: "Information in Information Theory is equivalent not to order, but to disorder in the sense that it takes more binary decisions — more ones and zeros, more bits of information — to describe disorderly situations or objects than orderly ones."

[6] Luenberger (2006) reviews current applications of Information Theory and Broadbridge and Guttmann (2009) discuss current concepts of entropy. Pfeifer (2006) describes some of the technical problems involved in applying Information Theory in biology and reviews examples of its inappropriate use. What apparently attracted people to applying Information Theory in biology was the way it quantifies information, something the qualitative approaches that deal with meaning and value are unable to do. Yockey (2005, p. ix), for instance, believes that Information Theory enables biology to become a quantitative and computational science that "can now take its place with theoretical physics without apology." Biology should not have to apologize, however, for the fact that the world of living organisms is largely governed by meaningful information, which is not amenable to the types of quantification that underpin the physical sciences.

References

Baddeley R, Hancock P, Foldiak P (2000) Information Theory and the Brain. Cambridge University Press, Cambridge

Bolton D, Hill J (2004) Mind, Meaning, and Mental Disorder: The Nature of Causal Explanation in Psychology and Psychiatry. Oxford University Press, New York

Broadbridge P, Guttmann AJ (2009) Concepts of entropy and their applications. Entropy 11(1): 59–61

Butler MH, Paton RC, Leng PH (1998) Informational processing in computational tissues. In M Holcombe & R Paton (Eds), Information Processing in Cells and Tissues. Plenum Press, New York

Cairns-Smith AG (1996) Evolving the Mind: On the Nature of Matter and the Origin of Consciousness. Cambridge University Press, Cambridge

Dreyfus H (1972) What Computers Can't Do: A Critique of Artificial Reason. Harper & Row, New York

Edelman GM (2004) Wider Than the Sky: The Phenomenal Gift of Consciousness. Yale University Press, New Haven

Forbes N (2004) The Imitation of Life: How Biology is Inspiring Computing. MIT Press, Cambridge

Glynn I (1999) An Anatomy of Thought: The Origin and Machinery of the Mind. Oxford University Press, New York

Gray J (2004) Consciousness: Creeping Up on the Hard Problem. Oxford University Press, Oxford

Hawkins J (2007) Why can't a computer be more like a brain? EEE Spectrum 44(4): 21–26

Hofstadter DR (1979) Godel, Escher, Bach: An Eternal Golden Braid. Basic Books, New York

Kåhre J (2002) The Mathematical Theory of Information. Kluwer Academic Publishers, Boston

Kelley AE (2005) Neurochemical networks encoding emotion and motivation. In J-M Fellous, MA Arbib (Eds), Who Needs Emotions? The Brain Meets the Robot. Oxford University Press, New York

Kosko B (2006) Noise. Penguin Books, New York

Küppers B-O (1990) Information and the Origin of Life. MIT Press, Cambridge

Luenberger DG (2006) Information Science. Princeton University Press, Princeton

Montague R (2006) Why Choose This Book? How We Make Decisions. Dutton (Penguin), New York

Pfeifer J (2006) The use of Information Theory in biology: Lessons from social insects. Biol Theory 1(3): 317–330

Picard R W (1997) Affective Computing. MIT Press, Cambridge

Pierce JR (1980) An Introduction to Information Theory: Symbols, Signals and Noise, 2nd edn. Dover Publications, New York

Sarkar S (2005) Molecular Models of Life: Philosophical Papers on Molecular Biology. MIT Press, Cambridge

Schneider ED, Sagan D (2006) Into the Cool: Energy Flow, Thermodynamics and Life. University of Chicago Press, Chicago

Searle JR (2004) Mind: A Brief Introduction. Oxford University Press, New York

Severin WJ, Tankard JW Jr (1979) Communication Theories: Origins, Methods, Uses. Hastings House, New York

Shannon C, Weaver W (1964) The Mathematical Theory of Communication. The University of Illinois Press, Urbana

Turing AM (1950) Computing machinery and intelligence. Mind 236: 433–460

Von Bayer HC (2004) Information: The New Language of Science. Harvard University Press, Cambridge

Von Neumann J (1958, 2000) The Computer and the Brain. Yale University Press, New Haven

Wicken JS (1987) Evolution, Thermodynamics, and Information: Extending the Darwinian Program. Oxford University Press, New York

Yockey HP (2005) Information Theory, Evolution, and The Origin of Life. Cambridge University Press, New York

Young P (1987) The Nature of Information. Praeger, New York

Author's Note

A book is an odd form of communication, since there is no actual contact between the sender and the receiver. We do not know each other, do not know where the other one is coming from, do not know what axes each of us has to grind, or what oxes each has to gore. There is no two-way connection, no dialogue, no feedback. It is like a message addressed "to whom it may concern" and launched blindly into space, hoping it will land somewhere. I think we would have a better chance of understanding each other if we had an opportunity to sit down and talk things over. You could ask me what I was trying to get at (and why) and tell me where I did not get things right or did not make sense, and I could learn where I have erred or stretched things a bit too far.

Especially in a book about new ideas, there is no way of guaranteeing that what seems so clear to the author will also be clear to the reader, either because the author has not captured his or her thoughts well enough or they do not fit the reader's frame of reference. As Robert Rosen observed in his preface to *Life Itself* (1991): "The problem was to compress a host of interlocking ideas, drawn from many sources, which coexist happily in my head, into a form coherently expressible in linear script. Moreover, there was the problem of trying to indicate the richness of many of the ideas, which in themselves want to ramify off in many directions, while keeping them focused on the primary problem." Ironically, there is no guarantee that a book about meaningful information actually conveys any.

So I thought we might try to make up for this by availing ourselves of a little technology. I have established a blog-site for anyone who wants to comment on the book or any of its parts, ask questions, point out errors and omissions, or add their own thoughts about any of the topics covered. Both favorable and unfavorable remarks are welcome. I will do my best to answer any entries directed specifically to me. The book consists of some beginning thoughts about meaningful information, not the last words about the topic—so there is a lot left to be done. I look forward to starting that conversation and hearing from you at:

http://meaningfulinformation.blogspot.com

A. Reading, *Meaningful Information: The Bridge Between Biology, Brain, and Behavior*, 151
SpringerBriefs in Biology 1, DOI 10.1007/978-1-4614-0158-2,
© Springer Science+Business Media, LLC 2011

Index

Matter and energy, 2, 3, 5, 9–11, 14, 20, 27,
     28, 33, 47, 48, 54, 58, 79–81, 111, 112,
     131, 144, 146
Meaningful discourse, 92
Meaningful information, defined, 1, 4, 9–13
Meaning, literal, 90–92
Memory, 3, 4, 19, 24, 28, 36, 48, 51, 56,
     64–69, 74, 140
Mental illnesses, 137, 139–141
Messages, 2–4, 6, 11, 13, 19, 28, 29, 51,
     83–88, 90–94, 105–108, 145, 147–149
Metaphor, 18, 92
Microscope, 98, 144
Mind and Brain, 20, 149
Mind's eye, 40
Mirror neuron, 42
Misunderstanding, 2, 92, 94, 136
Models
     flawed, 136
     mental, 57, 75–78, 90, 136, 137, 141
     representational, 58, 60
Mona Lisa, 29
Moral sentiments, 114
Morse Code, 88
Mortality, 127
Multicellular organisms, 51, 83, 97, 107

N
Natural Selection, 11, 19, 22, 23, 28,
     30, 37, 46, 50, 84, 85, 88, 105,
     108, 109, 112, 116, 124, 129, 135,
     136, 140, 145
Nature, 3, 6, 10, 11, 19, 25, 27–30, 39, 46,
     54, 56, 58, 81, 83–85, 87, 94, 97, 99,
     102, 106, 109, 113–115, 120, 127,
     132, 136, 145
Neocortex, 128, 140
Neural connectivity, 47, 63, 68, 76, 94
Neural representation, 32, 93–94
Neural signal, 35, 36, 47, 54
Neuron, 1, 20, 24, 31, 35, 36, 40, 42,
     46, 48, 59, 63, 67, 101–103, 126, 131,
     139, 149
Neuronal configurations, 13, 20, 42, 74, 132
Neuronal connectivity, 10, 19, 20, 45,
     51, 58, 73
Neurotransmitter, 46, 98, 101, 139
Newton, Isaac, 19
Noise, 11, 28, 33, 39, 56, 144, 148
Nonconscious activity, 128–129
Nonverbal communication, 93
Nucleotide, 106–108, 148, 149
Nurture, 11, 89, 90, 106, 109, 120

O
Object detection, 29, 30, 54
On/off signal, 35, 46, 84, 87, 91, 144, 147
Open Systems, 57, 61
Order, 3, 4, 10–14, 21–24, 30, 37, 55, 57, 58,
     60, 79, 84, 90, 99–101, 103, 106, 107,
     116, 118–120, 123, 131, 146, 149
Organization, 2, 5, 11, 12, 14, 73, 101,
     137, 147

P
Pain, 1, 29, 32, 37, 39, 60, 66, 112, 114–116,
     119, 120, 124, 126, 128
Parasympathetic nervous system, 39
Parts and wholes, 56–57
Pattern detection, 4, 10
Paul Revere, 84
Peirce, Charles Saunders, 4, 7, 48, 87
Perception, 19, 32–33, 35–42, 45, 55, 56, 60,
     63, 65, 67, 68, 70, 75, 112, 113, 123,
     125, 137, 139, 145
Perceptual Control Theory, 24
Perceptual discrimination, 36
Phenotype, 106
Pheromones, 37–38, 42
Placebo, 127
Plato, 27
Pleasure, 112, 113, 115, 116, 119, 124, 131
Predicting, 18, 23, 46, 77–79, 138
Predictions, inaccurate, 138–139
Prefrontal cortex, 139, 141
Printing press, 89
Processing levels, 127–128
Proteins, 11, 29, 37, 38, 46, 55, 67, 87,
     97–103, 106–109
Proteome, 103
Proteomics, 103

Q
Qualia, 113–114
Qualitative, 31, 54, 80–82, 113, 119, 149

R
Radar, 29, 144
Rain Man, 64
Reality, 2, 36, 76–81, 131
Reasoning, 53, 74–75, 81, 113, 118, 120, 125,
     128, 141, 145
     pattern-based, 74, 75
     rule-based, 74
Recall, 58, 60, 64–70, 125–127